I0891025

Social process and the city

URBAN STUDIES YEARBOOK 1

Social process and the city

EDITED BY PETER WILLIAMS

Routledge
Taylor & Francis Group
LONDON AND NEW YORK

First published in 1983 by
George Allen & Unwin Australia Pty Ltd

Published with the assistance of the
Urban Research Unit, Australian National
University

Published 2018 by Routledge
2 Park Square, Milton Park, Abingdon, Oxon OX14 4RN
605 Third Avenue, New York, NY 10017

Routledge is an imprint of the Taylor & Francis Group, an informa business

National Library of Australia
Cataloguing-in-Publication entry:

Social process and the city
Bibliography
ISBN 0 86861 237 5
ISBN 0 86861 245 6 (pbk.).
1. Sociology, Urban. I. Williams, Peter, 1947-.
(Series: Urban studies yearbook).
307.7'6

ISBN 13: 978-0-415-86037-6 (pbk)

Contents

Contributors

All communications regarding possible contributions should be addressed to the Editor, Dr Peter Williams, Urban Research Unit, Research School of Social Sciences, Australian National University, P.O. Box 4, Canberra, ACT 2605, Australia. Books for review should be sent to the Review Editor, Dr Peter Spearritt, School of History, Philosophy and Politics, Macquarie University, North Ryde.

CONTRIBUTORS

Carolyn Allport is Senior Tutor in the School of History, Philosophy and Politics at Macquarie University and co-author of Working Lives.
Ray Bunker is Head of the Department of Planning, South Australian Institute of Technology and author of Town and Country or City and Region?
Graeme Davison is Professor of History at Monash University and author of Marvellous Melbourne.
Robert Freestone is a PhD student at the Centre for Urban and Environmental studies at Macquarie University.
Elizabeth Harman is lecturer at the School of Social Inquiry, Murdoch University and co-editor of State, Capital and Resources in the North and West of Australia.
Evan Jones and Frank Stilwell are respectively Lecturer and Senior Lecturer in Economics, University of Sydney. Frank Stilwell is the author of Economic Crisis Cities and Regions.
Chris Maher is Senior Lecturer in Geography at Monash University and author of Australian Cities in Transition.
Brian McLoughlin is a consultant with C.E.S. Ltd. He is author of Urban and Regional Planning, a Systems Approach.
Chris Paris is Research Fellow in the Urban Research Unit, Australian National University and co-author of Not Much Improvement, Urban Renewal Policy in Birmingham.
Andrew Parkin is Lecturer in Politics at Flinders University and author of Governing the Cities.
Peter Saunders is Lecturer in Sociology at the School of Cultural and Community Studies, University of Sussex and author of Social Theory and the Urban Question.
Peter Spearritt is Senior Lecturer in Politics at Macquarie University and author of Sydney Since the Twenties.
Nigel Thrift is Senior Research Fellow in the Department of Human Geography at the Australian National University and co-author of Time, Spaces and Places, a chronogeographic perspective.
Peter Williams is Research Fellow in the Urban Research Unit, Australian National University and co-editor of Urban Political Economy and Social Theory.

ACKNOWLEDGEMENTS

In preparing this first volume of the Yearbook there are numerous people I'd like to thank for their assistance, encouragement and perseverance. Two of the staff of the Unit deserve particular mention, Susan Wells who prepared all of what became a difficult text with utmost efficiency and Kate Vasey who used her librarianship and literary skills to the full to impose some order on errant manuscripts. Pat Troy and other colleagues were quick to assist when needed as was Peter Spearritt. Launching a new venture requires considerable effort and I am also grateful to the Editorial Boards for their help and advice.

P.W. July 1982

Introduction Theory and process in urban studies

Peter Williams

A review of urban studies in 1982 reveals the wide spread of disciplines engaged in the analysis of the processes at work in urban areas. Without attempting to define urban studies, a developed interest in 'the urban' is now apparent in anthropology, economics, geography, history, politics and sociology as well as the sphere of professional activity, city planning. This suggests the importance given to an understanding of the ways social processes operate in cities and the increasing pressure for social sciences to address 'urban' problems. Often these appear deceptively simple but, as will be apparent through the pages of this volume, the processes at work in urban areas often defy any quick interpretations or solutions. We still have much to learn both about contemporary issues and processes and their historical antecedents.

It is this growing multi-disciplinary involvement and a continuing concern for the future of urban areas that has stimulated the introduction of the <u>Urban Studies Yearbook</u>. It is intended to provide an avenue for papers which contribute to current debates and issues in urban studies. Our hope is that the Yearbook will publish articles that cover a spectrum of interests: questions which are theoretical and conceptual in focus; matters requiring empirical examination and elucidation; and enquiries into contemporary policy debates. The papers will address issues which are considered to be of importance both in Australia and elsewhere. The aim throughout will be to assist the development of both Australian urban studies and urban studies in general. Moreover, underlying this intention is the view that Australian research can be as significant in this general context as research stemming from other countries.

Central to the above is the establishment of frameworks within which debate and analysis can be conducted on a cross-national basis. In this introductory essay,current developments in urban studies are discussed and an outline given of the conceptual issues being confronted. This provides the background against which the articles presented in this volume have been written. Each contribution is then briefly reviewed.

THEORY AND PROCESS

THEORY AND SOCIAL PROCESS

The growth of interest in urban studies has been paralleled by significant developments in the way urban questions have been tackled (Dunleavy, 1982). It is difficult to give a concise definition for the field, in part because we are still discovering interconnections and lacunae that require incorporation and elaboration. An illustration of this is the growing impact of feminist studies and the dawning realisation that the structure and functioning of cities has had a vital influence on gender relations. Similarly, many of the taken-for-granted aspects of state intervention, for example, planning as a reformist activity, are being challenged not simply through supply-side economics but also through a critical reappraisal of the nature of bureaucracies and organisations. As such urban studies, like society, is constantly changing and its subject matter is strongly conditioned by the concerns of the day.

Such immediacy should not however be allowed to overshadow two trends which are apparent in the longer term. The first is a concern with understanding and a recognition that specific assumptions and viewpoints are implicit in all forms of argument and analysis. There has emerged a considerable pressure for such assumptions to be made explicit and for such frameworks to become the subject of debate. This concern with theory has been seen by some as a retrograde step in that it shifts debate away from pressing issues into the more abstract and unreal world of concept-formation. But since our understanding of problems rests upon a recognition of the strengths and weaknesses of the approaches being adopted, it is essential that such work proceeds and becomes a recognised part of urban studies. It is very difficult to initiate effective change without understanding the problem and it is difficult to understand a problem without first considering the ways in which it is being defined. Such an argument for theory is not an argument for theoreticism, any more than the encouragement of empirical research necessarily leads to empiricism. Having said this, it is essential that theoretical positions are explored in some detail. A good illustration of this is the current debate being conducted as to the merits of weberian and marxist approaches. Although the argument is at times dense and abstract it is in part concerned with the question of causation, clearly a matter of some importance. There will be those who find these debates frustrating but they provide a means of linking urban questions to more general arguments within social science thus contributing to an erosion of unhelpful disciplinary boundaries. Moreover the argument is not for the abandonment of empirical research but rather a more balanced consideration of both theory and empirical work and a recognition of the value of both.

The second discernible trend is a concern with process and a dissatisfaction with the description of pattern. Recent work by urban historians reveals the complex processes of development which have contributed to the form and function of urban areas.

Awareness of change and continuity over time - and through space - is essential to any grasp of the functioning of cities. Related to that, is the question of what sort of process? The work of sociologists, geographers, political scientists and economists demonstrates the very different ways of interpreting the processes at work and the sharp conflicts which exist between approaches. The history of urban studies has been one of compartmentalised debates within disciplines which have ignored or denied relevant contributions from other disciplines. Such positions are now more widely challenged and it is apparent from recent work that, although perhaps starting within a disciplinary perspective, very few urban analysts can now retain a narrow disciplinary focus and adequately address contemporary issues. Cross-disciplinary transfers hold out the prospect of a genuinely inter-disciplinary debate. This carries with it the possibility that by increasing our understanding of urban areas as 'crucibles' in which social processes form and change, we will be better placed to make the interventions so obviously needed.

There is not room in this introductory essay to review in detail the different frameworks which have been advanced in explanation of the way urban areas function. Crudely they have operated at three levels - individual, institutional and societal with each posing different but related questions about the processes at work. Presently the emphasis has shifted towards societal explanations, reflecting amongst other things, the ways the fortunes of particular cities and their inhabitants are so clearly linked to changes in national and international economies. However, although the so-called political economy approach is currently ascendant there is now a general recognition that there have been many problems with its application - not least of which are an over emphasis on economic processes and functional relations and a neglect of the contradictory forces at work. Social reality is not easily collapsed into single dimensional processes. The new critical engagement in urban studies requires that such matters be given careful attention.

Urban studies in Australia : towards an Australian urban studies?

One problem too rarely confronted in urban studies is the question of the place-boundedness of particular discussions of theory or process. Often the very complexity of the argument acts to hide the way the very specific experiences of one area are drawn into a general model (for example, as in Castells' book, The Urban Question) which others may uncritically seek to apply. It is plainly apparent from most university texts that many place-specific questions are transposed between countries without modification or even comment. This 'homogenising' process can then act against the development of a body of research and analysis which is sensitive to local variation. Szelenyi (1981) suggests this has happened in Australian sociology and examples can be found elsewhere in the social sciences. The pressure to conform and to replicate rather than to contrast and challenge leads both to bad theory and a weak grasp of process. One of the strengths of the new interest displayed by historians is the challenge such work can pose to prevailing universal orthodoxies about urban areas. At the same time,it is necessary to guard against the slide into a chaotic world of case studies with no generalising principles - thus the need for balance.

The emergence of a more self-critical urban studies is a matter of some importance. So too is the question of urban studies within a particular country. Given a focus on application to contemporary problems, it is obviously important that the work being conducted is not insensitive to local issues. Similarly, the need for greater specificity within the frameworks being applied offers valuable opportunities for localised research to both enhance our understanding of particular areas and strengthen theory. Indeed, the argument here would suggest that, a body of work in one country, integrated with urban studies in general through a concern with particular debates and issues, could act as an important and powerful lever upon that wider literature. Moreover a concern with theory as well as empirical research will allow us to isolate and explain features specific to Australian society. In this way, Australian urban studies might avoid the fate Szelenyi accords to Australian sociology (1981:44):

> Australian sociology, for example, has made very little in the way of a genuine attempt for authentically original Australian research. There is nothing new about the theories which have so far been developed to account for the growth of Australian society or the manner in which it presently functions. For the most part they are little more than rather boring versions of what has already been discovered, explained, demystified elsewhere. Sociologically, Australia is America uncovered on a smaller less dramatic scale.'

The present situation in Australia is promising in a number of respects. A new critical engagement in urban studies is underway as reflected in a number of books published in recent years. Despite the relatively small number of scholars involved, say in comparison to Britain or the United States, there is a growing cross-disciplinary involvement in urban studies (eg, Stretton's Ideas for Australian Cities) reflecting the important role that cities in particular, and urban development in general, have played in the growth of Australia. Economists, economic historians, historians, political scientists and sociologists (as well as geographers and planners) have all begun to display considerable interest in urban questions as Peter Spearritt's review later in this volume indicates. Despite claims of a 'natural myopia' regarding urban studies (Parkin, 1982) cross-disciplinary activity has been in evidence for some years - witness the urban concerns of Butlin's Investment in

Australian Economic Development or Mathews' Public Investment in Australia. Since the founding of the Australian Institute of Urban Studies in 1967 there has also existed a forum for cross-disciplinary exchanges in the pursuit of better policy, a rare achievement in itself, while a number of academic institutions have made an explicit commitment in this area by funding posts and a range of activities.

Of all the different attributes to consider perhaps the most important lies in the rich possibilities which exist for research. The development of Australia through separate colonies into a federated system with three tiers of government raises innumerable questions regarding the state and forms of intervention around economic and social questions. The role of overseas capital, the absence of a fuedal land-owning class, the key role played by urban development in fixed capital formation and a whole range of other questions related to issues such as housing, immigration, urban politics and regional growth offer an interesting and important research agenda. The similarities and contrasts between Australia and say Britain and the United States hold out the prospect of a range of important comparative work. In addition, the close links between Australia and the Pacific region, between developed and developing areas, suggests it is an ideal place in which to study the linkages between first and third world urbanisation.

SOCIAL PROCESS AND THE CITY

A number of themes have been outlined in this short introduction; the focus on theory and process, a conscious recognition of spatial and temporal variation and a stress on debate and evaluation. In this first issue we have taken up these themes and developed them with regard to the processes operating in urban areas. Thus although the substantive examples will vary between papers the overriding concern in each is to develop an understanding of the ways urban areas function and an awareness of the different interpretations which might be applied.

A recurring theme in urban studies relates to the definition of 'the urban' (Dunleavy, 1982). What is an 'urban process', an 'urban problem' or 'urban sociology'? While many would agree with Pahl (1975) that the city is merely an arena in which general social processes are played out there remain doubts as to whether such definitions solve the problem or merely avoid it. Can we have a body of work called urban studies if we cannot define what urban is? Such a question is not merely academic since in seeking to answer it one is beginning to define the issues as well as the boundaries for the urban problems which command so much attention. Dunleavy (ibid:2) argues that we should

> 'renounce any connection between particular spatial locations (cities and towns) and urban studies. Instead we

> should simply define 'urban' processes as certain types of socio-economic activity which can be studied as a connected field of research wherever they are located in geographical space. If we choose the right kind of social process to focus on then we may well find that the results of our research are especially helpful in understanding what goes on in cities or towns, but this is something which needs to be empirically examined.'

While some will have sympathy for his argument this still begs the question of what kind of process. Moreover it is ironic that Giddens, a leading social theorist, has recently argued that 'the city cannot be regarded as merely incidental to social theory but belongs at its very core' (Giddens, 1982:140). Clearly, no easy definition will suffice and Dunleavy is right to stress that the subject matter cannot be assumed in advanced and that much further exploration is required. In the opening article Evan Jones and Frank Stilwell take up this issue. Using examples from contemporary debates about the city they seek to explore the different dimensions within any single 'urban' problem. Although they give considerable attention to the societal forces at work, they also consider what practical solutions can be offered. In doing so, they also confront those who consider that a structural analysis is incapable of addressing the question of reform.

Their concerns are reflected in Brian McLoughlin's article on planning education. He focusses upon the way planning education and planning theory have ignored the social context in which planning takes place. Drawing upon his background both as a proponent of the systems approach to planning and a key figure in planning education, Brian McLoughlin's critique of the current situation in Britain asks how urban problems are defined and researched and what sorts of solutions are proposed.

Peter Saunders' contribution on weberian social theory casts this debate into an explicitly theoretical realm. With his book Social Theory and the Urban Question, he offered a summary and critique of existing theory and only in the final chapter did he explicitly elaborate his own position. In this article we are given a much fuller account of Saunder's position vis a vis Marx and Weber and his challenge to marxist analysists. Later in this volume, Chris Paris reviews the work of Saunders and other writers on urban politics and this can be usefully counterposed with some of the arguments advanced here.

This concern with the nature of the forces at work in urban areas is extended both empirically and conceptually in the articles which follow. Carolyn Allport looks closely at the post-war planning process and the provision of social infrastructure. Her concern throughout is to expose the way that the development of the built environment had the effect of constraining the role women might play in post-war society. The failure to provide adequate bus-services and child-care facilities, for example, restricted the opportunities women had for paid work. At the same time the inadequacies of the suburbs as centres of everyday life were, in large part, only overcome by the un-paid labour of women. Recognition of the gender bias of much research takes us part way towards a necessary reconstitution of urban studies and a better awareness of the so-called structural forces mentioned. At the same time there is the danger that, in adapting macro-scale theories of urban change, generalisations may obscure very real and vitally important local variations. Chris Maher's article on population migration looks at the very different rates of change being experienced in urban centres across Australia. While he does not fully resolve the question of how to explain these differences he is able to identify a number of the components which must be taken into account.

Questions surrounding the role of the state, resource development and their linkages with metropolitan growth and change have become matters of considerable interest in forging theoretical links as well as addressing practical problems. Elizabeth Harman addresses these with a specific focus on Perth. The exploitation of Australia's mineral resources and the benefits presumed to flow from this are major political issues. Harman's analysis suggests that the local benefits relative to costs are easily overstated and that despite the degree of assistance given by the state government many of the most significant benefits accrue to enterprises outside of Western Australia and indeed, outside of Australia.

There has been a widespread failure to extend urban studies beyond the limited concerns of industrial and commercial development, the housing and labour markets, and journey to work questions. Such limited approaches receive a sharp rebuke through Graeme Davison's article on urban reform. Like Carolyn Allport, he exposes the social processes underlying our taken-for-granted world. He draws the linkages between concerns with the health and vitality of children living in working-class districts, middle-class urban reform movements and the kindergarten movement. He thus spans debates about the impact of urbanization, the role of the colonies as providers of a new healthy British race, and inter-class relations in the city. Moreover, as Davison himself argues, although the urban reform movements have been subjected to considerable criticism, much of our current urban infrastructure of schools, halls and kindergartens can be traced back to this period.

A different historical treatment is given in Robert Freestone's review and bibliography of Australian urban planning 1888-1948. This is the most comprehensive treatment published on Australia and it is a timely contribution to a world-wide upsurge of interest in planning history. Freestone's review of the literature reveals the many gaps in our present understanding of the role of planning and he argues that, in part, these can only be remedied by adopting a more theoretical approach to the subject.

Following the main body of articles come two further sections. First the Policy Report which each year will give a brief summary of developments in one or more of the States. In this first issue we publish Ray Bunker's report on South Australia. Despite the withdrawal of the Federal Government from explicit intervention in urban affairs, Bunker shows that policy initiatives are still on the agenda in at least one state. In the Reviews section we begin with an overview of published work on Australian urban studies. Peter Spearritt finds much that is lacking, both with respect to academic contributions and the role the publishers have played. Chris Paris' review article on urban politics focusses our attention on one of the emergent areas of debate in contemporary urban studies. He offers a valuable critique of Peter Saunders <u>Urban Politics</u> and Patrick Dunleavy's <u>Urban Political Analysis</u> and then uses Power , Wettenhall and Halligan's <u>Local Government System of Australia</u> to develop an agenda for urban political analysis in Australia. Andrew Parkin's discussion of Lloyd and Troy's <u>Innovation and Reaction</u> is a complement to this. An alternative perspective on process is given in Nigel Thrift's review of recent urban atlases, raising the question of how far such documents can proceed beyond description.

Throughout the volume, the focus is upon processes at work in cities and the debates that surround them. Urban studies is no longer the narrow prerogative of geographers and planners and the very breadth of interest in urban areas is likely to maintain some of the present dynamism. Similarly, the revived concern with theory and explanation should ensure a much more critical engagement in research and analysis. Moreover, it is apparent that the most important questions being asked at present are those which focus upon the inter-connections between different realms, between regional development and the city, politics and the economy and so on - a process well illustrated both in this volume and recent books such as Hall (1982) and Harloe and Lebas (1981). In following this path it becomes increasingly difficult to retain a single disciplinary interest, raising the real posibility of growing levels of integration within the social sciences. All these different trends give grounds for great optimism regarding urban studies, both with regard to its analytical capabilities and its capacity to confront contemporary issues.

REFERENCES

Dunleavy, P., (1982) Perspectives on Urban Studies, in Blowers, A., Brook, C., Dunleavy, P., and McDowell, L., eds., <u>Urban Change and Conflict; an interdisciplinary reader</u>. London, Harper and Row.

Giddens, A. (1982) <u>A Contemporary Critique of Historical Materialism</u>, London, Macmillan.

Hall, P. ed. (1982) <u>The Inner City in Context</u>. London, Heinemann.

Harloe, M and Lebas, E. eds., (1981) <u>City, Class and Capital</u>, Arnold, London.

Pahl, R., (1975) <u>Whose City</u>, Harmondsworth, Penguin, paperback

edition.
Parkin, A., (1982) Governing the Cities; the Australian experience in perspective. Melbourne, Macmillan.
Saunders, P., (1981) Social Theory and the Urban Question, Hutchinson, London.
Szelenyi, I., (1981) Comments in an interview conducted by Bryan Mellonie, Quadrant, April:43-47.

1 When is an urban problem not an urban problem?

Evan Jones and Frank Stilwell

It is widely recognised that cities are characterised by a range of problems: atmospheric pollution, long journeys to work, congestion in public facilities and poor access to these facilities in many areas, sub-standard housing, pockets of high unemployment, rising incidence of crime, and so on. Some analysts talk of an "urban crisis" (Stilwell, 1979). What are the origins of these problems? One possible consequence of an examination of such problems from a specialist discipline of 'urban studies' is the attribution of their cause to the character of urban structure itself. The issue of causation is vital. To use Alonso's phrasing, it is tempting to interpret all problems manifest in cities as problems of cities (Alonso, 1971). However, it is important to ask whether the problems arise from spatial considerations, for example, city size, the spatial structure of the cities or inadequate planning of urban development. Or are the problems due to other causes and their appearance in spatial form merely a manifestation of these less obvious causes? Can we posit distinctive effects due to (a) the effects of the inherited environments, both physical and human-made, (b) the imperatives of the mode of production, (c) the impact of the state apparatus and (d) the role of space *per se*? In general, is the spatial character of 'urban and regional' problems causal or symptomatic?

The issue is of some considerable significance. First it raises questions of academic importance. The study of social phenomena within the last one hundred years has been characterised by substantial compartmentalisation into academic 'disciplines'. Although this compartmentalisation is understandable in terms of attempts to develop a partial understanding with specific analytical tools, the process has been accentuated by traditional academic structures. New generations of would be intellectuals are schooled within a pre-existent disciplinary structure, and many successful researchers stay within an inherited analytical framework without making any conscious decision as to its methodological advantages. For example, although urban and regional economics has developed separately from its 'parent', orthodox economic theory (because of the lack of a spatial dimension to the latter), its vision, analysis and emphasis have remained influenced by the vision, analysis and emphasis of orthodox economics. Traditional academic structures also make it more likely that an urban economist will have had more exposure to economics than to urban sociology, for example. Such disciplinary isolation tends to produce specific analyses that may be rigorous on their own terms, but also potentially distortive of the 'global' picture. The problem is perhaps most acute for the geographers, who naturally emphasise spatial considerations. Fortunately, the last ten years has witnessed a variety of attempts to build a more comprehensive analysis of spatial forms (for example, Harvey, 1973, 1976 and 1978; Castells, 1977, 1978; Tabb and Sawers, 1978; Harloe, 1977; Carney, Hudson and Lewis, 1980; Mullins, 1981). Our contention is that an interdisciplinary approach is not only desirable, but essential if "urban problems" are to be systematically analysed.

Secondly, the issue is important because of its vital implications for public policy. The emphasis on spatial causes of urban problems tends to provoke calls for remedial policies such as decentralisation, improved urban planning, and so on. However, if the problems are simply manifestations of more fundamental factors, then such remedial policies do not attack the cause: different and/or more drastic responses are needed in conjunction with spatial policies. This perspective would suggest that "urban problems" can only be resolved through a more general assault on the socio-economic system which generates social inequalities, inefficiencies of resource allocation and utilisation, and environmental decay. As a consequence, policies which are appropriate for urban problems may range from the most superficial to the most profound. Indeed, it is not inconceivable that some problems may be deemed insoluble.

The practical significance of these policy considerations in the Australian context is obvious: the last Labor Federal Government sought to redress social inequality through the spatial policies of the Department of Urban and Regional Development with quite limited success (Lloyd and Troy, 1981). What a reformist party could - or should - attempt on coming to office again is a matter deserving the most careful consideration and requiring a coherent analysis which the last Labor government lacked. The approach of this paper gives priority to these implications for policy: thus, the degree of difficulty involved in remedial 'policies' is a dominant consideration in the grouping or ordering of the factors.

It should be stressed that what follows is not intended as a definitive and unique work. We are not concerned here with the development of a conceptual structure adequate for an holistic analysis of urban problems. Neither do we claim complete originality. The issue of causality has even been formally acknowledged by the applied neo-classical economists who now generally acknowledge the importance of the distinction between problems <u>of</u> cities (that is, caused by spatial factors) and problems <u>in</u> cities (that is, having non-spatial causes). Such writers use the existence of the latter to support the critique of government intervention to limit city size (for example, Richardson, 1973). Marxists have also confronted the question more formally, predictably locating the roots of urban problems in the nature of capitalism as a mode of production. Our concern is to clarify the issue of causality by some simple propositions: (a) to posit the concept of a 'locus' of causation - there exist several readily discernible loci of causation of urban problems; (b) that the most adequate analysis of causation would place urban problems as predominantly non-spatial in origin and spatial only in manifestation; (c) that the non-spatial causes of urban problems, though related, are relatively autonomous in that they cannot be reduced to a single cause, and each presents its attendant degree of difficulty for an appropriate public policy. In addition, the use of Australian examples should highlight the significance of the issues in a specific context.

We begin by examining historical and geographical influences which shape the "built environment" and impose severe limits on public policies. Secondly, we study the influence of the mode of production on urban problems. Though the policy implications arising from this approach are, in general, revolutionary rather than reformist, we indicate some areas in which there is scope for improvement within the structural constraints imposed by capitalism. Thirdly, we examine certain aspects of the state apparatus which have influence on urban problems, and discuss whether particular problems stemming from the nature of the bureaucracy transcend the constraints imposed by the mode of production. Finally, we consider the influence of spatial form _per se_.

GEOGRAPHICAL AND HISTORICAL INFLUENCES

Geographical influences impose major constraints on the pattern of urban and regional development. Indeed, the Australian physical environment probably provides the ideal supporting example. The lack of adequate and reliable rainfall, the limited distribution of arable soil, and the impediments to communication created by the Great Dividing Range have set significant limits on development. This 'determinist' position was forcefully represented earlier in the century by a geographer, Griffith Taylor, whose persistent arguments against an optimistic view of Australia's future akin to the possibilities, say, of the United States, gained him little gratitude. Taylor believed that "the best economic programme for a country to follow has in large part been decided by Nature. Man is able to accelerate, slow or stop the progress of a country's development. But he should not, if he is wise, depart from the directions as indicated by the natural environment" (Taylor, 1940). With the benefit of hindsight, it is clear that Taylor understated the case of the extent to which political-economic factors may influence the structure of development. Changes in agricultural technology have opened up more possibilities of rural development. Even more dramatically, the extraction of Australia's abundant mineral resources has led to the development of new settlements in various areas, most notably in the north-western region of Western Australia. Nevertheless, the restrictions of soil, climate and topography continue to provide a physical basis for the lack of intensive agriculture and the extra-ordinary concentration of population and economic activities evident in Auatralia.

Historically, geographical considerations have interacted with political-economic considerations in shaping the spatial distribution of population and economic activity. The high degree of urbanisation in Australia - and its concentration in the six state capitals in particular - is closely linked to the colonial relationship with Britain. Separate colonies were established for reasons of defence, isolation of penal colonies and the need to control free settlement. Within each of the individual colonies, public administration was and is highly centralised. The focus on the major ports as the trans-shipment points for the export of primary goods and the import of labour and other commodities further reinforced the tendencies towards centralisation. In a sense, the general unevenness in the Australian economy which was the outcome of colonial dependence can be seen to have manifested itself spatially in a striking unevenness in the development of urban and rural areas. This duality in the economic structure - and in the spatial structure - is a general feature of dependent Third World economies, and Australia's colonial origins have stamped its spatial structure partly in this mould. Indeed, particular features of the Australian situation, such as the absence of any significant pre-capitalist rural peasantry and the ability to exploit the resources of the interior through radial rail networks, provided additional reasons for the dominance of the capital city within each colony. Simultaneously, of course, the separate development of the individual colonies prevented the growth of metropolitan primacy in the continent as a whole.

Having developed for this assortment of historical reasons, the capital cities came to have cumulative advantages, as they attracted investment in manufacturing and service activities, and developed a wide range of commercial, financial and social facilities. New businesses - other than those based on natural resources - have typically preferred a metropolitan location because of the large labour market, access to market and ancillary services. Moreover, the immigrant groups - especially those from Southern Europe (who accounted for about half of the population growth since World War II) - have tended to be disproportionately concentrated in the cities, seeking proximity to jobs in the manufacturing sector and social contact with their own ethnic groups (Collins, 1975). Thus the forces of "circular and cumulative causation" have tended to reinforce the high degree of metropolitan dominance.

It should be stressed that the last decade has seen something of a reversal of this process as Sydney and Melbourne in particular have failed to secure the share of population growth experienced during the "long boom" of the 1950s and 1960s. Explanations for this vary and might include the effect of the structural change in the economy towards mineral extraction and processing and away from city based manufacturing industries (Stilwell, 1980) as well as the relative attractiveness of the non-metropolitan environment (Lansdown, 1977). The changing balance between the State capitals in their shares of population and employment growth is also striking. Nevertheless, these changes in the pattern of urban and regional development occur at the margin, and for all the cities - including those whose share is no longer growing - the historically determined structure imposes great constraints on the process of capital accumulation.

The overwhelming feature of the built environment is its inertia (Harvey, 1978). Whatever the causal influences operating historically, the vast investment in urban development, once undertaken, make for a great "stickiness" in the workings of the space economy. Of course, the constraints are not absolute. The private sector of the economy certainly does not treat them as such. Restructuring of the built environment is occurring constantly as the present interests of investors conflict with the concretised 'past' interests of previous investors. We are witness to the continual replacement of shops/residences/offices in and around the central business district by buildings of greater density; the segmentation of inner-city communities and destruction of parklands by intra-urban freeways; the closure of manufacturing business in the inner-city resulting from movements of capital to the outer suburbs and overseas; and the disapperance of whole towns subsequent to the closure of key mining or manufacturing businesses. Lamarche (1976) argues that urban restructuring is a crucial element in creating the conditions for capital accumulation in that it helps reduce the circulation time of capital (the average time period for which capital is tied up in a particular project). Yet the temporal limits to this process of spatial restructuring necessary for capital accumulation impose important constraints on the pace and the form of capital accumulation. In Harvey's words, "capitalist investment has to negotiate a knife-edge path between preserving the exchange value of past capital investments in the built environment and destroying the value of those investments in order to open up fresh room for accumulation." (Harvey, 1978).

Policy Implications

Changing the physical environment is a matter of technical expertise, adequate finance and collective will. Endowed with copious quantities of the last resource, the Israelis, at least for the time being, have transformed their country. The Chinese are outstanding in their attempts to make productive use of their water resources. Australian governments have made intermittent attempts to transcend the imposing frontiers facing viable settlement, for example, in the Snowy Mountain project and the Ord River Scheme, although such projects are notable for their rarity as much as for their success. The ravages of leaching, drought and flood continue to impede rural decentralisation. Given also an apparent cultural tradition which is antagonistic to a role for government which is strongly interventionist, far-sighted and coordinated, these sorts of geographical influences impose substantial obstacles to remedial public policy.

Similarly, the built environment - shaped by the historical needs of capital and constrained by geographical factors - is both the object of public policy and a constraint on its operation. As we have seen, the interests of profit are constrained by the built environment: the constraints on public policy are that much stronger because of the potential conflicts both with the inertia of the built environment (partly embodying the interests of past investment) and with the interests of contemporary private investment.

Decentralisation policy, which has attracted a long-standing commitment in Australia (in rhetoric if not in action), provides an obvious example. Steering growth away from the major capital cities has been an objective accepted by all the major political parties, and has led to the development of an array of public policies, particularly in the South-Eastern States (Neutze, 1978:Ch.4, Stilwell, 1974:Ch.13). Most of these policies involve direct subsidies to capital such as low-interest loans for the purchase of land and the construction of factories, rail-freight subsidies, subsidies for training of workers, payroll tax rebates, and so on. However, it is a strategy that has operated "against the wind", in that the perceived advantages of a metropolitan location have remained dominant for most industries. Each of the six State capitals continued to capture a rising share of their respective State population (see figures in Stilwell, 1974:17). It is true that the cumulative growth of the major cities has slowed in the past decade but, as previously noted, this is at least partly the result of structural change in the economy towards an emphasis on the mineral sector. There have also been major changes in the immigration flow and the rate of natural increase (Borrie, 1978) which have slowed metropolitan growth but which cannot be considered to be directly contributable to the effects of regional policy. The very fact that the major urban areas contain the majority of the national population makes it likely that they will have to accommodate the majority of population growth (on the conservative assumption that fertility rates are similar between city and country and that immigration is distributed between city and country in rough proportion to the existing population). The normal tendency would be for both public and private investment to accommodate to these trends in respect to the provision of housing and social services. Thus, the cumulative tendencies are perpetrated. In effect, past factors, "concretised" in existing spatial structures, make public policies harder to effect.

More fundamentally, in so far as the cumulative forces compound the centralised spatial structure which developed in part because of Australia's historical colonial subordination, policies seeking to alter the adverse impact of 'the past' must eradicate any continuing impact of the same influences. That is, the nature of Australia's contemporary relations with overseas interests must be considered explicitly in terms of their continuing impact on the built environment. Urban policy needs to be formulated in conjunction with policies facilitating more control over international trade and investment. Unless, for example, further restrictions including locational restrictions, are placed on multinational corporations, the government lacks the necessary degree of control over the pattern of development. Similarly, an indirect effect of a general reduction of tariffs would be to lessen the ability of the government to determine the pattern of regional development.

INFLUENCES ON THE MODE OF PRODUCTION

Massey has argued that the so-called 'regional problem' is not fundamentally a spatial phenomenon, but derives from the restructuring of capital associated with periods of economic crisis. (Massey, 1979, Massey and Meegan, 1979). More generally, the possibility exists that urban problems may be due to the structural imperatives of the socio-economic system in which they arise. The mode of production may play a central role in the determination of the form of urban development and in its attendant "problems". The state of development of the forces of production - including the state of technology - constrains the technical possibilities for handling the requirements of large cities, for example, in respect of transport. Equally, the social relations of production - the private ownership of the means of production and the associated class system - impose constraints on the form of urban development. The point can be most easily illustrated by a number of examples. Four are selected: social inequality, unemployment, housing and transport, these being four of the most commonly discussed urban problems. The discussion is meant to be no more than illustrative.

Social inequality in cities is one aspect which has been subjected to considerable analysis (for example, Stretton, 1970, Sandercock, 1975, Troy, 1981). Inequality of income is fundamental to the capitalist system. Profits are both a source of income to the capitalist class and the incentive for further investment. The workers, on the other hand, derive income from the sale of their labour power, and these wages are a cost of production. The relative levels of profits and wages are an outcome of an antagonistic relationship. Within the wage-earning categories there are also major inequalities of income which reflect historical differences in the ability of different occupational groups to use their power at the expense of less well organised groups. These inequalities take an obvious form in cities in that different income groups are associated with distinctive patterns of residential segregation. Thus spatial inequality is in large part the spatial manifestation of an inequality systematically generated by the mode of production. The upper income groups are able to some extent to buy their way out of urban problems by purchasing housing in areas such as Sydney's eastern suburbs, with low levels of atmospheric pollution, good transport facilities and so on. This highlights the fact that urban ' problems' are often not problems for all, but problems for the relatively disadvantaged sections of the community. The labelling of certain social problems as "urban" therby diverts attention from their markedly uneven incidence within the urban area.

Unemployment provides a second example. Conceptually, unemployment is one means by which social inequality is manifested and reproduced. But it is a means of vital significance, and deserves to be singled out, not merely because of its importance in assisting capital accumulation through the dampening impact of the 'industrial reserve army' on labour's bargaining power, but also because of its profoundly adverse impact on the well-being of those involved. Unemployment appears as a spatial problem for the same reason as 'social inequality' because unemployment is concentrated in particular occupational groups whose residences are spatially concentrated. Thus, concentration of unemployment is a permanent feature of the spatial landscape. In addition, the mobility and restructuring of capital in response to economic crises both accentuates spatial concentration and imparts a dynamic instability to employment in particular areas. Such capital mobility tends to generate pockets of high unemployment, and the experience of depressed regions in the United Kingdom, Europe and the United States is a clear reminder that capital ultimately has no regional loyalties. In the Australian case, the current period of structural change has particularly significant employment implications for the cities. The manufacturing sector has been concentrated in the south east of the nation in the horse-shoe-shaped arrangement of cities from Newcastle, through Sydney, Wollongong, the towns of the La Trobe Valley, Melbourne, Geelong, Adelaide, Port Pirie, Port Augusta and Whyalla. To a certain extent, capital has been mobile within the manufacturing sector, producing structural unemployment arising from changed requirements in skills and location of employment (Bureau of Industry Economics, 1981). With the switch of capital investment from manufacturing to mineral-extraction, and processing, with the offshore relocation of some of the manufacturing activities, and with the 'rationalisation' and introduction of capital-intensive technology in many of the remaining plants, the employment base of many of these areas has been eroded. Further, given that <u>within</u> the major cities working class people dependent on these sorts of jobs are clearly spatially concentrated, the outcome has been high concentrations of unemployment in particular localities (Stilwell, 1980:Ch. 7).

A third illustration involves housing. A recent paper by Mullins (1981) shows how the character of housing in Australian cities is related to the history of Australian capitalism. The heavy emphasis on private home ownership is attributed in part to the relative absence (particularly in the capital cities other than Sydney and Melbourne) of an industrial capitalist phase intervening between the phases of mercantile and corporate capitalism. Other explanations have also been posited (for example, Kemeny, 1981). Perhaps what is of most contemporary significance is the problem of the high cost of housing in Australian cities which derives in large measure from the private character of the housing market. The private nature of land and housing ownership permits speculative holdings which add significantly to the cost of accommodation. The history of urban development in Australia is rife with instances of speculative activity on a grand scale (Sandercock, 1979:Ch. 2) and this phenomenon gives rise to many problems. It inhibits access of low income groups to the housing market and thereby intensifies the problem of poverty. Moreover, as Sandercock argues (1979:Ch. 1), it undermines the possibility of effective metropolitan planning and the efficient provision of public services as well as adding significantly to land prices and thence to the general inflationary problem. Even Edward Heath, when Conservative party Prime Minister in the UK, conceded that specualtion shows "the unacceptable face of capitalism". It is a key source of wealth for the owners of property while having nothing directly to do with any productive contributions.

Urban transport provides a fourth example. The petrol-driven automobile is an instrument of great personal convenience, albeit an extraordinarily expensive one. However, there are social costs, including urban congestion, air pollution, and the segmentation of inner-city communities by the freeways necessary to carry the traffic. The private automobile serves the profit interests of the automobile, oil and tyre industries, and a host of ancillary industries, but it is far from being the best form of urban transport in terms of cost and health. In part, this situation arises because of the problem of uncoordinated individual decisions: car usage may be individually rational (especially where public transport is poor) but the sum of those individually rational decisions leads to a collectively irrational outcome (Edel, 1973). Second, there is the role of large corporations in structuring the set of transport alternatives in a manner which serves their own profitability at the expense of public amenity; the destruction of the public transport system in Los Angles by a consortium headed by General Motors, Standard Oil of California and Firestone Tyres is the most commonly cited example (Snell, 1974). Finally, there is the role of the state in providing road transport infrastructure at public expense, thereby cheapening relatively the most socially irrational mode of transport and effectively subsidising the large corporations who profit from it (Lacey, 1976).

Policy Implications

To the extent that spatial problems arise as a consequence of the profit orientation of a capitalist society, it would appear that they are soluble only if the socio-economic system is transformed. Confronted with this demanding ultimatum, and given that a non-capitalist Australia is not on the immediate agenda, there is some point in examining more closely the nature of the above set of problems. Are these sorts of problems associated with fundamental features of capitalism or do they arise from specific features which might conceivably be altered through policies of reform? The latter possibility should not be underestimated because capitalism has shown itself in the past to be adaptable within certain limits in the face of problems which have the potential to generate challenges to its political and social stability. Our answer is that some problems are indeed potentially amenable to modification within the framework of the capitalist system.

The degree of speculative activity is one example. As we have argued, this is a major element in the high cost of housing. In a capitalist society, markets for the sale and re-sale of commodities (including housing) are necessary, of course, and one might argue that speculation arises simply because of the existence of such markets coupled with uncertainty about the future pattern of urban development and price levels. But speculation per se does not create value in the marxist sense, in that it is associated with exchange rather than production of commodities. It is an appropriation of value already embodied in existing goods - housing in this case. From this point of view, speculation is not fundamental to the viability of the mode of production. Of course, one should not underestimate the political and institutional impediments to dealing with speculation; large-scale speculative activity is an integral part of urban development, as Sandercock has shown in the case of the State Government of Victoria. Widespread dwelling ownership in Australia, coupled with the onerous nature of mortgage finance, locks a substantial number of middle-income people into support for the 'right' to appropriate capital gains on land and housing. The Menzies government apparently understood quite well the positive impact to polical conservatism that home ownership could exert (Kilmartin and Thorns, 1978). It is because of these impediments that radical proposals like the single land tax of Henry George have few proponents now and current proposals for capital-gains taxation typically exclude gains arising from owner-occupied property. However, it is not utopian to argue that the scope of speculative activity can be reduced. The ambitious interventions of the state government in South Australia have been successful in keeping down the price of land and housing, though the inter-state variations in demand and supply conditions need to be acknowledged (Sandercock, 1979:Ch. 5). The tentative steps of the Labor federal government in 1972-75 to introduce Land Commissions was another example of policy to stabilise urban land prices, albeit not living up to the hopes and intentions of the designers of the policy (Troy, 1978).

Urban transport provides another example. A wide range of

urban transport systems function in existing capitalist societies, from the extreme case of auto-freeway 'hegemony' in Los Angeles to the convenient public underground railways of Paris and London. The history of urban transport in Australia is a mixed one. The turn-of-the century surface radial railways were an advanced and efficient development and they still carry millions of passengers weekly. The aesthetic and non-polluting tram-way network of Melbourne is an international showpiece. Yet these developments were compatible with the contemporary interests of business and speculators (Butlin, Barnard and Pincus, 1982). Since World War II and mass automobile ownership, public transport has received scant attention or funds from State governments. Paradoxically, the recent community lobbying for more extensive and effective public transport has burgeoned as a consequence of the urban sprawl made possible firstly by rail network and later by widespread automobile ownership. The phenomenal expansion of Sydney's frontiers and the onerous rise in automobile costs helped put a Labor government into power in New South Wales and there has been continuing pressure for improved standards in public transport.

THE NATURE OF THE BUREAUCRACY

Certain aspects of urban problems are not explicable simply in terms of the economic system of capitalism. The most obvious illustration of this is the intensity of many of the same sorts of problems in socialist cities. Various explanations are plausible *a priori*, including the role of the bureaucracy, be it within a capitalist or socialist state. It must be acknowledged that analysis of this issue is particularly difficult because it involves general questions about the role of the state in a capitalist economy and specific questions about the role of the bureaucracy within the state apparatus. Perhaps the most modest approach is to begin by looking at the working of the bureaucracy in practice and the criticisms commonly levelled at it. We are then in a better position to assess whether the problems are inherent in the bureaucracy because of its role within the capitalist system, or whether they derive from particular features of the bureaucracy which are amenable to reform, at least in principle. Thus, we confine our remarks in this section to certain aspects of Australian urban problems which seem to be attributable to the nature of the bureaucracy within the state sector.

There has been a long standing commitment in the Australian case to some form of urban planning. Its major manifestation, at the local level as in many other countries, is the zoning of non-compatible urban land uses and the attempt to use zoning as a means of restricting uncontrolled development, particularly on the urban periphery. In practice, this has had limited effectiveness, as such restrictions have been modified in the face of pressure from property-owning groups - witness the failure to effectively implement the scheme of the Cumberland

County Council for the future of Sydney and the erosion of the Green Belt (Stretton, 1970:Ch. 9). Worse, such zoning has often created the conditions for speculative profits as property owners stand to make vast windfall gains as the result of re-zoning of land from rural to urban use (sometimes with the benefit of inside information from the planning bodies themselves). Again, the history of Melbourne's development provides a classic example of the integration of urban "planning" and the interests of local capital (Sandercock, 1979:Ch. 3 and 4).

Added to this general problem of the role of vested interests in urban planning are some specific problems associated with the nature of the planning process. The problem of co-ordination between different elements in the planning process is one and liberal writers on urban policy have made much of these issues (for example, Lawrence, 1972, Neutze, 1978). In part, the problem can be traced to difficulties arising from the federal system. There exists predictable impediments to smooth coordination between the national, State and local governments. An example of federal-state conflict is housing policy, as manifest in the tensions underlying the 1973 Commonwealth State Housing Agreement (Carter, 1980). The impediment of State rights to comprehensive national policies is as relevant in housing as it is in other areas. In part, the problem of coordination also arises in the priorities of different, and sometimes competing, instrumentalities, both at the Federal and State levels. At the Federal level, for example, the lack of effective integration of the Department of Urban and Regional Development and the Cities Commission inhibited urban policy (Wilson, 1978). At the State level, major housing developments have been undertaken - including complexes of public authority housing - in outer suburbs such as Sydney's Green Valley and Mount Druitt. That a substantial volume of dwellings has been erected since the 1930s and that there was much good intent behind the developments cannot be denied. However, these developments have often been poorly synchronised with the provision of other facilities such as schools, social amenities, transport and employment opportunities.

A second set of criticisms relates to the long-established tendency in the 'planning' institutions to focus on <u>physical</u> solutions to socio-economic problems - a tendency based on what has been terms the "facility-centred theory of social change" (Sandercock, 1975). Historically, this has been manifest in programmes of slum clearance and in attempts to provide adequate sewerage and drainage in the metropolitan areas. Currently, it is associated with the development of physically attractive suburban areas which conceal rather than ameliorate the socio-economic problems of the lower-income groups - indeed this may even reinforce such problems. The result is the growth of peripheral communities of low-income people with no local employment, poor access to the rest of the metropolis and high concentrations of personal and social problems (Bryson and Thompson, 1972).

Additional features of the bureaucracy which are not conducive to the amelioration of urban problems involve particular attitudes. Elitism is one and the bureaucracy is often seen to adopt a paternalist approach towards social problems associated with the lower income groups. As an example, public housing is said to cater for the housing needs of an insufficient percentage of poor households because they are too much trouble - "they do not pay rent, they are dirty, they demoralize their neighbours" (Priorities Review Staff, 1975). The major beneficiaries of public housing are the 'deserving poor' - those whose styles of living are considered acceptable by the relevant authorities. It may also be observed that the operations of the bureaucracy show a fear of the unexpected and the original that comes with rigid hierarchies and a vulnerability to criticism; and a detachment that the public servant sees as rational in the interests of long-term personal stability.

A more fundamental question also needs to be posed in order to determine the assumptions underlying the role of the bureaucracy. Is there a commitment to planning at the national level, involving foresight, co-ordination and control of the allocation of resources so as to achieve whatever goals are deemed desirable - such as natioanl self-sufficiency, economic growth, equity, environmental quality and community development? Or is there merely a grudging willingness to engage in <u>ad hoc</u> intervention in an economy directed predominantly by private sector decisions - in effect, an ideological aversion to systematic planning. The Australian case is more readily interpreted in terms of the latter approach (McFarlane, 1968).

How can these characteristics of the public service be explained? To a certain extent the term 'public service' is itself a misnomer. Many individuals join the public service for self-interested reasons - the public service becomes a means to secure employment and career advancement as in other occupations. Indeed, historically, many individuals of Roman Catholic background were compelled to seek employment in the Federal public service, being denied opportunities in the private sector (Crisp, 1976). Yet the choice of the public sector for secure employment does not explain why the outcome of public service operations should be misdirected. That has more to do with the lack of a cultural tradition of 'public service, which itself has to be explained.

The absence of a commitment to systematic planning and intervention demands special consideration. This state of affairs has not always been a permanent feature of political life in Australia such that it could be ascribed as innate in the Australian character. Colonial governments organized the structure of the economy - they arranged capital flows, facilitated migration and dispensed land; they financed and built urban infrastructure. By 1930, this substantial involvement in the structure and direction of the economy was being gradually curtailed (Butlin <u>et al</u>, 1982). It seems plausible that this major historical role of the public sector in Australia was due to the absence of a local capitalist class of sufficient size and unity to finance and coordinate structural development. The last sixty years has witnessed the growth of an urban capitalist class centred in the large financial and manufacturing concerns, which has sought and achieved a more direct role in determining the country's economic priorities (Cochrane, 1980). The role for government becomes a subordinate one, and this is manifest tacitly in the policies of the Liberal Party. To a certain degree, the Labor Party in office has attempted a more interventionist role - the Chifley Government's attempt to nationalize the trading banks was a major manifestation in the economic shere of a broader vision. A similar vision is implied in many of the interventionist policies of the Whitlam government. Of relevance here, the Department of Urban and Regional Development was created as an explicit commitment to more comprehensive planning in the pursuit of redistributive objectives.

Yet the post-war initiative has resided generally with the Liberal-Country Party coalition, and its view towards planning and intervention has presided over post-war development. Within the Federal bureaucracy, the Treasury has been the major embodiment of this anti-interventionalist philosophy. It is the Treasury which has become the 'policeman' for the permanent enforcement of this philosophy, regardless of developments in public opinion or of which party holds office. Thus the mild interventionism recommended by the Report of the Vernon Committee in the 1960s was strongly opposed by Treasury economists. The Treasury also became the 'bete noir' of the Whitlam government's ambitions for mineral development and for social policies effected through the Depaprtment of Urban and Regional Development.

The major contribution of the Treasury to directing the nature and form of Australia's economic priorities is poorly understood and inadequately researched. The Australian Treasury is derivative of the British Treasury and its traditional close ties with the interests of banking capital. There is a strong conservatism on fiscal matters, linked to the Treasury's role as the nation's purse-keeper (a conservatism which varied little during the so-called liberal age of Keynesianism). In addition the dominant ideology reflects a purist free-market interpretation of economic life, and this forms the 'intellectual' basis for the antagonism to planning in any form. The immediate reason for the attachment to this ideology is apparent - departments like the Treasury and the Industries Assistance Commission are staffed with individuals trained by the tertiary institutions in the rigid orthodoxy of neo-classical economics.

One is drawn to infer that some of the problems arising from the bureaucracy within the state arise from the dominant capitalist values. Moreover, the apparent inefficiencies of the bureaucracy may exist because it is handed responsibility in areas which are either not-profitable or have suffered because of profit-taking, while simultaneously being denied the revenues to operate in these areas effectively. For example, the continual difficulties of the State Housing Commissions in Australia arise from the persistent and substantial demand for public housing brought on by the pincer effect of low wages and high housing costs.

There are two qualifications to this these which links the bureaucracy to the structural imperatives of a capitalist economy. Firstly, a state bureaucracy inevitably develops a certain autonomy with interests of its own, particularly in its own reproduction, and in the assertion of its power in specific decisions. The bureaucracy cannot be said to serve the interests of the capitalist class in any structurally deterministic manner. It is also an arena of class struggle and, indeed, itself changes the nature of the class structure. A self-serving bureaucracy is one manifestation of this situation, and secrecy, caution and its special brand of elitism are manifestations of this relative autonomy. These problems are not specific to bureaucracies within the capitalist state (Konrad and Szelenyi, 1979).

Secondly, there is not one unified capitalist 'class', but a variety of 'fractions' of that class, spanning numerous dimensions - sectoral (for example, manufacturing and rural); small versus monopoly capital; domestic versus overseas capital and so on. Different fractions will exert varying degrees of influence on the bureaucracy at different times in each capitalist country. In Australia, the continuing failure to develop a tradition of national and regional economic planning acts to support overseas investors to the substantial detriment of both the interests of domestic investors and the Australian working class. By contrast, the coordinating activities of the state in capitalist countries like France, Germany and Japan tend to reflect more a nationalist pragmatism which transcends free-market ideology but acts ultimately in the interest of domestic capital accumulation (Shonfield, 1965, Halliday, 1975).

Policy Implications

An alteration of the structure and/or behaviour of one part of the state requires intervention either from another part of the state or pressure from outside the state apparatus. The former possibility raises the question of relationship between the government and the public service. Can an elected government effect significant reforms to structure and operations of the bureaucracy? The experience of the Labor government in Australia between 1972 and 1975, though disappointing in so many respects, suggests that there is some scope for significant changes of this sort. That government set about gathering information on potential failings and how they might be overcome - by the establishment of the Coombs inquiry into the Federal bureaucracy and sending missions overseas to gain perspective on alternatives. Acknowledging the inertia in the existing bureaucracy, it also established a major new department, the Department of Urban and Regional Development, and staffed it with individuals whose values and perceptions were already sympathetic in most cases (Lloyd and Troy, 1981). This was a means of effecting policies, through the bureaucracy, which would contribute to the amelioration of urban problems. While it is true that there were many problems such as the lack of a fully comprehensive long-term strategy and conflicts with other departments, the Department of Urban and Regional Development could eventually have provided the basis for the creation of a system of regional economic planning, given continuing support from successive governments. This is not to infer that creating new departments is a sufficient condition for overcoming the problems of elitism and economic conservatism discussed earlier. It is just one method. Another involves the movement of key individuals from senior posts on the advent of a reformist government in essence a "night of the long knives". These are most difficult issues but the point is that changes are possible within the state apparatus and that these may go some way towards alleviating bureaucratic impediments (Wilenski, 1980).

Outside the state apparatus, pressure groups can also influence decisions about the form of urban development and thus reduce the influence of an unreconstructed bureacracy. The Builders' Labourers' Federation's "green bans" influenced the face of Sydney in the interests of aesthetics and leisure. The Sydney group 'Shelter' coordinates less priviledged groups in activities oriented towards more adequate housing. Environmental groups such as the New South Wales Total Environmental Centre, through persistent pressure, can exert a long-term influence on decisions affecting the physical environment. The bureaucracy is not autonomous, it may have considerable inertia but it operates in a political environment which reflects the balance of class forces. Thus, even though one acknowledges the general nature of the role of the bureaucracy within the capitalist state, one can observe elements both in other aspects of the state apparatus and the society at large which can constrain its impact on urban development.

THE INFLUENCE OF SPATIAL FORM PER SE

Is spatial form per se responsible for urban problems? The general tenor of the argument in this paper is that spatial structure is predominantly a manifestation of problems which are due to more fundamental causes. Having down-graded the causal significance of spatial form, one must necessarily acknowledge two particular ways in which it remains significant.

Firstly, physical space is an essential medium through which more profound causes operate - class structure in Sydney is translated into locational patterns characterised by water views for the wealthy and hot, barren plains for the lower wage-earners. More formally, we can say that the spatial form is one dimension of the set of social, economic, political and cultural interrelationships. It is the arena in which these interrelationships operate. As one writer has recently put it in more pejorative terms:

> What appears as a thing-like collection of physical properties and attributes - the arrangement of workplaces and households, transit routes moving volumes of goods and people, cultural monuments and humble shelters, differentiated areas of productivity and poverty, scenic sites filled with appreciative travellers, accessible sites providing public services, kitchens and bedrooms, deteriorated inner cities and repetitive suburbs, more or less "developed" regions, fortunate sites for raw materials, strategic locations, administrative boundaries, defensible territories and turfs - is in essence a socially created field of relationships between individuals and classes congealed and concretised to maintain and perpetrate the exploitative dominance of those who own and control the means of production" (Soja, 1981).

The interaction of space and social processes is the key point. If one ignores the social processes then the analysis focuses on things to the exclusion of human interrelationships (what marxists call "commodity fetishism"). On the other hand, ignoring the mediating function of space leads to the sort of accusations made by Massey (1973) in criticising orthodox economics because it treats economic activity as if it takes place "on the head of a pin".

Secondly, physical space <u>compounds</u> problems originating in more fundamental causes. In this second sense, spatial form can be said to be a cause of urban problems. The compounding influence of spatial form on social inequality provides a major example. The key variable is the standard of social service provision. This is typically lower in low-income areas, as has been demonstrated in recent studies of Australian cities (Troy, 1981). Where services are provided by local government, spatial inequalities are compounded because of the greater financial difficulties of local government in areas of low socio-economic status (with a correspondingly low revenue base) (Manning, 1973). Educational inequalities are also particularly important, because differential socialisation in the schools reduces the extent of inter-generational social mobility (Bowles and Gintis, 1976). The simple fact of distance also inhibits interaction and information about "how the other half lives", leading to a greater complacency towards inequality than might otherwise pertain. In these various ways, the spatial form which inequalities take provides a basis for the further entrenchment of those inequalities.

<u>Policy Implications</u>

These sorts of issues were a prime concern of the Federal Labor government in the period 1972-75. Policies such as the area improvement programme (which involved payments to local governments in deprived areas to help in the improvement of public facilities) did make some contribution to the reduction of spatial inequality. The purchase of a significant area of housing in Glebe did help to maintain low-income housing in one part of inner Sydney during a period of rapid gentrification and associated inflation of housing prices and rents. Certainly the impediments to redistributional policies are great but, to the extent that urban problems arise from the compounding effects of urban residential segregation, they are amenable to some amelioration through public policy. Moreover, unless one is prepared to argue (as the current Federal Government does) that there is a "natural" limit to the share of wages (vis a vis profits) in the national income, then there is no general reason to suppose that redistributive policies per se threaten the viability of the capitalist system. As we have earlier argued, inequality is an essential feature of the system, but urban policy could, in principle, ameliorate the degree of inequality.

Catley and McFarlane (1974) have claimed that by focussing on inequalities associated with location, the Labor government (and the Department of Urban and Regional Development in particular) diverted attention away from the more fundamental origins of inequality in the sphere of production. However, there is no reason in principle why a redistributive spatial policy should not be a complement to, rather than a substitute for, policies towards ownership, control and taxation of wealth. To illustrate the point, the establishment of state-owned enterprises to provide employment in urban areas with high rates of unemployment is more consistent with reformist objectives than subsidies to employers in the private sector. In respect of housing, direct subsidies to low-income groups can be expected to be distributionally more effective than general policies to encourage home-ownership through taxation concessions (Carter, 1980, Kemeny, 1981). Also, as mentioned earlier, policies to control speculation in urban land and housing can be expected to alleviate inflation and to prevent the accumulation of unearned income; by contrast, physical restrictions on peripheral urban expansion - for example, green belts - tend to fuel the inflationary problem and to increase the income of the owners of existing metropolitan land. All these examples are no more than illustrative. The general point is the crucial need to integrate policies towards urban form with policies which embody more general social and economic goals, and particularly those concerned with distribution of income and wealth.

CONCLUSIONS

Many "urban problems" are spatial manifestations of more profound causes. We have identified three principal categories - the inherited environment, both physical and man-made; the imperatives of the mode of production; and the nature of the bureaucracy and associated ideologies, part of which derives from the economic imperatives of the mode of production and part of which is autonomous of these imperatives.

If this thesis is substantially correct, a major implication is that urban problems become much more difficult to ameliorate than is commonly supposed. Each of the three major areas is relatively immovable. The inherited environment is either unchangeable (the physical Australia can never be converted into a physical United States) or adaptable only at considerable expense. The demands of profit and the maintenance of the social relations of production in a capitalist society are persistent, to say the least. Bureaucracies are entrenched and ideologies permeate mass consciousness to the detriment even of those who hold them. Revolutions against vested interests and ideas are rare.

This is a pessimistic appraisal but it is more realistic than the typical discussion of public policy. The eradication of many urban problems, as we have shown, requires dramatic solutions. Nevertheless there is also some room for reformist policies directed at spatial structure and towards eliminating the compounding influence which spatial form itself can exert.

In the case of Australia, there has been a withdrawal from the concern with urban problems which characterised the early 1970s - both in terms of involvement by the Federal government and by popular protest movements such as resident action groups and the "green-bans". Meanwhile, many of the problems have intensified, for example, rapidly rising costs of transport for outer-suburban commuters, the problem of low-income groups seeking access to a very inflationary housing market, and the problems of unemployment concentrated in urban working class communities. There is a strong need to reintegrate the concern with production issues (technology, work, wages) and consumption issues (housing, transport) so as to confront the common causes of these problems. In the current circumstances of economic crisis, structural change and urban dislocation, the need for an integrated system of national, regional and urban planning becomes all the more imperative.

REFERENCES

Alonso, W. (1971) 'Problems, Purposes and Implicit Policies for a National Strategy of Urbanisation', Paper prepared for the National Commission on Population Growth and the American Future Institute of Urban and Regional Development, Working Paper 158, Berkeley, UCLA.

Badcock, B.A. (1973) The Residential Structure of Metropolitan Sydney,Australian Geographical Studies. 11, 1:1-27.

Borrie, W.D. (1978) 'Population Trends and Policies', in Scott, P., ed., Australian Cities and Public Policy. Melbourne, Georgian House.

Bowles, S. and Gintis, H. (1976) Schooling in Capitalist America: Educational Reform and the Contradictions of Economic Life. London, Routledge and Kegan Paul.

Bryson, L. and Thompson, F. (1972) An Australian Newtown, Harmondsworth, Penguin.

Bureau of Industry Economics, (1981) The Structure of Australian Industry - Past Development and Future Trends, Research Report No. 8, Canberra, AGPS.
Butlin, N.G., Barnard, A. and Pincus, J.J., eds. (1982) Government and Capitalism: Public and Private Choice in Twentieth Century Australia. Sydney, Allen and Unwin.
Carter, R.A. (1980) 'Housing Policy in the 1970s', in Scotton, R.B. and Ferber, H., eds., Public Expenditure and Social Policy in Australia, Volume II, The First Fraser Years 1976-78, Melbourne, Longman Cheshire.
Carney, J., Hudson, R. and Lewis, J., eds., (1980) Regions in Crisis. London, Croom Helm.
Castells, M. (1978) City, Class and Power. London, Macmillan.
Catley, R. and McFarlane, B. (1974) From Tweedledum to Tweedledee. Sydney, ANZ Book Co.
Cochrane, P. (1980) Industrialisation and Dependence: Australia's Road to Economic Development. St. Lucia, University of Queensland Press.
Collins, J.H. (1975) 'The Political Economy of Post-War Immigration', in Wheelwright, E.L. and Buckley, K., eds., Essays in the Political Economy of Australian Capitalism. Vol. 1, Sydney, ANZ Book Co.
Crisp, L.F. (1975) 'Politics and the Commonwealth Public Service', in Spann, R.N. and Curnow, G.R., eds., Public Policy and Administration in Australia: A Reader. Sydney, Wiley.
Edel, M. (1973) Economies and the Environment. Englewood Cliffs, Prentice Hall.
Halliday, J. (1975) A Political History of Japanese Capitalism. New York, Monthly Review.
Harloe, M., ed., (1977) Captive Cities. London, Wiley.
Harvey, D. (1973) 'Labour, Capital and Class Struggle around the Built Environment in Advanced Capitalist Societies', Politics and Society, 6:265-295.
Harvey, D. (1973) Social Justice and the City. London, Edward Arnold.
Harvey, D. (1978) 'The Urban Process Under Capitalism: A framework for analysis'. International Journal of Urban and Regional Research, 2, 1:101-131.
Kemeny, J. (1981) The Myth of Home Ownership: Private versus Public Choices in Housing Tenure. London, Routledge and Kegan Paul.
Kilmartin, L. and Thorns, D. (1978) Cities Unlimited. Sydney, Allen and Unwin.
Konrad, G. and Szelenyi, I. (1979) The Intellectuals on the Road to Class Power: A Sociological Study of the Role of the Intelligentsia in Socialism. New York, Harcourt Brace Jovanovich.
Lacey, G. (1976) 'The Dominant Car', Australian Left Review, 5th March:8-15.
Lamarche, F. (1976) 'Property Development and the Economic Foundation of the Urban Question', in Pickvance, C., ed. Urban Sociology. London, Methuen.
Lansdown, R.B. (1977) 'Issues of National Settlement Policy', The Australian Quarterly, 49:1, 56-62.
Lawrence, R.J. (1972) 'Social Welfare and Urban Growth', in Parker,

R.S. and Troy, P.N., eds., The Politics of Urban Growth. Canberra, ANU Press.
Lloyd, C. and Troy, P. (1981) Innovation and Change. Sydney, Allen and Unwin.
Massey, D. (1973) 'Towards a Critique of Industrial Location Theory', Antipode, December, 5, 3:33-39.
Massey, D. (1979) 'In What Sense A Regional Problem?' Regional Studies, 13, 2:233-243.
Massey, D. and Meegan R.A. (1979) The Geography of Industrial Reorganisation, Progress in Planning. 10:3. Oxford, Pergamon Press.
Manning, I. (1973) Municipal Finance and Income Distribution in Sydney. Urban Research Unit, Canberra, Australian National University.
McFarlane, B. (1968) Economic Policy in Australia: The Case for Reform. Sydney, Cheshire.
Mullins, P. (1981) 'Theoretical Perspective on Australian Urbanisation: 1. Material Components in the Reproduction of Australian Labour Power', The Australian & New Zealand Journal of Sociology, 17, 1:65-76.
Neutze, G.M. (1978) Australian Urban Policy. Sydney, Allen and Unwin.
Priorities Review Staff (1975) Report on Housing. Canberra, AGPS.
Richardson, H. (1973) The Economics of City Size. Farnborough, Saxon House.
Sandercock, L. (1975) Cities for Sale. Melbourne, Melbourne University Press.
Sandercock, L. (1979) The Land Racket. Canberra, Silverfish.
Shonfield, A. (1965) Modern Capitalism. London, Oxford University Press.
Snell, B. (1974) American Ground Transport. Sub-Committee on Anti-Trust and Monopoly, Washington, DC, US Senate.
Soja, E.W. (1981) 'A Materialist Interpretation of Spatiality', paper presented at the Workshop on the Geographical Transfer of Value, Department of Human Geography, Canberra, Australian National University.
Stilwell, F.J.B. (1974) Australian Urban and Regional Development. Sydney, ANZ Book Co.
Stilwell, F.J.B. (1979) 'An Urban Crisis?', Current Affairs Bulletin, 56, 7:22-31.
Stilwell, F.J.B. (1980) Economic Crisis, Cities and Regions. Sydney, Pergamon Press.
Stilwell, F.J.B. and Hardwick, J.M. (1973) 'Social Inequality in Australian Cities', The Australian Quarterly, 45, 4:18-36.
Stretton, H. (1970) Ideas for Australian Cities. Adelaide, Georgian House.
Tabb, W. and Sawers, L., eds., (1978) Marxism and the Metropolis. New York, Oxford University Press.
Taylor, G. (1940) Australia: A Study of Warm Environments and their effect on British Settlement. London, Methuen.
Troy, P.N. (1978) A Fair Price: The Land Commission Program

1972-1977. Sydney, Hale & Iremonger.
Troy, P.N., ed., (1981) Equity in the City, Sydney, Allen and Unwin.
Wilenski, P. (1980) 'The Left and the State Bureaucracy', The Australian Quarterly, 52, 4:398-414.
Wilson, R.K. (1978) 'Urban and Regional Policy', in Scotton, R.B. and Ferber, H., eds., Public Expenditures and Social Policy in Australia, Volume 1, The Whitlam Years 1972-75. Melbourne, Longman Cheshire.

2 The crisis in British planning education and research

J. Brian McLoughlin

It is tempting to see the causes of the crisis in planning education and research lying in the context of Britain's economic plight. As we are well aware, the world-wide recession is hitting Britain hard and some would say that the present government's economic policies are making matters worse. In addition to their general conduct of economic (or should one say "monetary"?) affairs, the government has made deep cuts in public expenditure affecting the local governments, who employ the vast majority of planners and the universities and polytechnics who conduct most planning research and education. All of this is being done in an atmosphere where "planning" of all kinds is distinctly unpopular not only among politicians but with the public at large.

I will argue that these manifestations are not the cause of the crisis I shall discuss. They certainly exacerbate the problem but they are not its roots. If Mrs. Thatcher were to resign tomorrow and public expenditure on planning, education and research were to be trebled next week, the crisis would remain. That is because it is an internal crisis, a crisis which is intellectual and fundamental rather than circumstantial.

For various reasons British planning practice, education and research have been drifting apart and losing touch and comprehension with each other. This is a serious state of affairs in any field of human thought and practice but it is especially serious in an applied or vocational area like planning. The crisis brought about by such a state of affairs can easily result in mindless rote learning in place of true education, of practice which is bereft of any theoretical basis, critique or explanation and theory which is innocent of the material reality of social practice.

Ideally, we should consider the three-way relationship between practice, education and research and study their dynamics over, say, the last twenty or thirty years. However to attempt that would be very difficult in terms of exposition and so, in the hope of greater clarity, I shall deal with them in turn and, in conclusion, draw them together once again.

PLANNING PRACTICE

British planning practice has a number of characteristics which are unique and which are important to remember in any analysis. First of all, it is clearly defined in statute and administrative law and applies (with modifications in Scotland and Northern Ireland) uniformly over the whole territory.(1) Secondly it is universal, that is, planning law applies to every square metre of land and buildings. Plans of various kinds cover the whole country and in all but a few cases development or change in land use requires explicit permission from the local planning authority. Thirdly, the system is operated from day to day by democratically elected local authorities with the central government acting as general coordinator and arbiter. Finally, like many other aspects of British public life, the service is highly professionalised, membership of the Royal Town Planning Institute being generally a pre-requisite for appointment and promotion beyond the lowliest ranks of the service, which itself is organised in powerful vertically disposed departments of the local authorities (McLoughlin, 1973).

It is against this background that we must understand the development of planning practice since its post-war inception in the 1947 Town and Country Planning Act. For the next twenty years or so - British practice was essentially the control or regulation of land use and development by means of the physical "development plan" and the exercise of "development control". Such practice assumed a large proportion of all developments would be for the public-sector and paid little attention to the underlying social and economic forces that set in motion development pressures. It also paid scant heed to the growth in private transport or to the role of the private sector. Lastly it assumed a broad social consensus and believed that professional town planners knew how to serve that consensus by physical arrangements of land uses.

By the latter part of the 1960s all that had changed. The Buchanan Report had put integrated land use and transport planning high on the agenda. The role of the private sector was now crucial and predominant. The importance of socio-economic analysis was recognised and social consensus could no longer be taken for granted. New stategic plans and policies were called for which in turn would be further specified in more detailed physical plans. The new planning system was explored in various "sub regional" studies, outlined by the work of the Planning Advisory Group (Ministry of Housing and Local Government, 1965), codified in the 1968 Town and Country Planning Act and given its theoretical underpinning by the so-called "systems approach" and associated procedural theories and analytical methods.(2) British planning became very broad, and aimed for "rationality" and "comprehensiveness". However the new dawn was to be short-lived. The economic crisis began in the mid 70s and "rational" analysis gave way to political pragmatism, wide horizons were narrowed by competing and parallel policy systems, public participation in structure plans was discredited and the statutory rigidity of the whole system failed to produce many timely or sharp edged policies (Drake, McLoughlin, Thompson and Thornley 1975). In 1978, ten years after the inception of the new system, there was still not a complete coverage of England and Wales by the rural structure plans whilst Scotland had gone its own more effective way into "Regional Reports".

By 1981, British planning practice was so diversified as to defy brief description. Variations in local social and economic conditions, battles between one locality and another for resources (or the ability to survive), battles between central and local government, between departments, professions and services, have resulted in a bewildering variety of styles of practice. Mrs. Thatcher's administration is dismantling all kinds of "strategic" planning but is only finishing off a sick patient anyway. Local planning is about everything from economic development to tree planting, from development control to entrepreneurial flurries to "get things done". Market forces are to dominate and official pronouncements indicate that development will normally be permitted. Planning has regressed to detailed local physical planning of a minimal kind. It is suspected by the political right of fouling up the workings of the market and by the left for helping to sustain the system by "papering over the cracks".

PLANNING EDUCATION

Over a similar period considerable changes have occurred in planning education. The 1950 Schuster Report noted with approval the predominance of postgraduate training for architects, engineers and surveyors but felt that there should be more openings for other disciplines within the planning profession.(3) It believed that planning was not so much a discipline as a complex field of application for a wide variety of skills combined in a team effort. The Report noted with interest but a distinct lack of enthusiasm that the undergraduate courses at Newcastle (Durham) and Manchester had begun in 1945 and 1949 respectively. The predominance of post-graduate education lasted until well into the 1960s but the growing output of undergraduate planners and the rising proportion of geographers, economists and other social scientists forced the profession to reconsider its views on planning qualifications. In 1963 and 1964 the ruling body of the Royal Town Planning Institute drew up proposals to liberalise entry to the profession. They were baffled by the vehemence with which their ideas were rejected and by the decisiveness - 3 to 1 - of their sacking at a special general meeting in January of 1965. For the time being, the idea of a single-skill profession ruled and the expansion of the University schools following the Robbins Committee Report, and later of the new Polytechnics, was very much biased towards undergraduate education (Cherry, 1974).

And yet both undergraduate and postgraduate courses - covered by the Royal Town Planning Institute's "recognition" conditions - moved towards a fairly uniform model of content (Kantorowich, 1967). This was based on the notion of a unique body of skill based on a distinct area of knowledge. Both were closely associated with procedural theories of planning and with the technical armoury of mathematical models and other quantified techniques. Little emphasis was given to substantive matters such as housing, transport, industry or commerce, to social issues such as urban poverty or regional inequality, or to the more general matters of the operations of the land market, the development process and the planning system.

Following the huge surge in demand produced by the local government reorganisations of 1974-75, output peaked in the late 1970s since when there has been a marked fall-off in demand and a struggle for survival by almost every planning school. These attempts to survive take many forms but common features include mid-career training, short courses, commissioned research, courses for school teachers, children and community groups. There is also a noticeable tendency for a large proportion of schools to offer a broad all-round education in the hope that this will recruit the student numbers on whom survival often depends.

PLANNING RESEARCH

Here, we do not need to go back to 1950 because planning research hardly existed until the foundation of the Social Science Research Council in 1965 and the Centre for Environmental Studies in 1967, (Llewelyn-Davies, 1967) (the former is threatened by savage budget cuts, the latter was closed in 1980 by the Minister for the Environment, Michael Heseltine). By far the largest source of research funds however came from the Department of the Environment which was formed in 1970. During the last decade, funds for planning research have grown spectacularly in relative terms but still form a small proportion of social science efforts and a trivial proportion of all "civil research and development" spending. The funding of planning research had hardly got under way when, in 1971, the "Rothschild principle" was announced whereby almost all government-sponsored research had to follow the so-called "customer-contractor" principle. That is, policy divisions of government - the "contractors" defined problems of interest to ministers and sought for academic "customers" to pursue the investigation. The implications for the levels of funds available for free-ranging critical studies of longer-term or strategic questions are obvious. It was left to the Research Councils and private foundations to support such fundamental work. They have not only been hampered by shortages of funds - now very serious - but also by other factors affecting planning research in particular.(4)

The Research Councils operate largely in "responsive mode"and very rarely take initiatives themselves. Unfortunately academics in planning schools have a very weak research tradition.

Planning educators, especially the more senior ones, more often than not come from a practice background and have encouraged consultancy rather than academic research. Planning schools often lack a research "atmosphere", and the few research students or assistants are therfore isolated and often on insecure contracts. Lively active researchers, usually the more junior tenured staff, lack senior leadership. Planning education is often very classroom and workshop-intensive which leaves little time for scholarly inquiry.

By contrast, much research which has been of value to planning practice and education has come from social science departments such as geography, economics, sociology and political science or from specialised centres, institutes and "think-tanks" such as the Centre for Urban and Regional Studies at Birmingham, the School for Advanced Urban Studies at Bristol and the late lamented Centre for Environmental Studies in London.(5) One of the most searching analyses of British planning, Peter Hall's monumental work, was carried out from geography departments and funded by a private foundation (Hall, Thomas, Gracey and Drewett, 1973)

Research which has thrown light on the problems of British cities and regions has usually been done outside planning schools. Research carried out in planning schools has been minute in scale, discontinuous, poorly funded, lacking in leadership and carried out by underpaid insecure junior workers. It has lacked a career structure, continuity and critical development. It has had an "adjunct" or ancilliary status to teaching and consultancy rather than being seem as a mainstream activity in its own right.

Academic critics have regarded much planning research as being either blandly descriptive of practice or critical in uninformed and unconstructive ways. It has dealt with minutiae and failed to grasp the underlying dynamics of urban and regional change. It has favoured procedural and technical matters and ignored substantive issues and problems. Practitioners have accused research for failing to study day-to-day practice and for being too abstruse, theoretical and jargon-ridden. (6)

As we move into the 1980s planning research is in a parlous condition. Public expenditure cuts bite deeply into funds. Only studies which are of immediate practical use to ministers will be funded by government and fundamental work will have to be supported by the Social Science Research Council on a drastically reduced budget. Academics are preoccupied with a battle to survive and research comes rather low in their immediate priorities.(7)

WHAT HAS GONE WRONG? - AND WHY?

A healthy vocational field should have rich linkages between education, research and practice. Ideally its practice would be based on a solid but evolving body of theory. Education would have a broad context but be internally specialised in various aspects of practice. Research would be the crucial link offering detailed critical explanations of practice on which to base both practice and education. Such an ideal is approached by other fields like medicine, aeronautical engineering and nature conservation. Why does British planning fall so far short?

The answer must be based upon an examination of the particular circumstances and characteristics of British practice and academic life. The distinctive feature which I believe offers the maximum explanation is the institutional character of British planning and in particular the ideology of professionalism which has been used to further the specific developments in practice, in education and research. The 1947 Act legitimised a struggle by town planners to carve out a niche in the bureaucratic power structure. This aimed at the creation of separate planning departments in the public service, each with its own chief officer. In order to achieve that, it was important to claim that planning was a distinctive operation based on unique knowledge and operational skills. This was done by tightening up the entry qualifications to the professional institute, obtaining status through a Royal Charter, negotiating "closed shop" arrangements with public sector employers, and the system of "recognising" courses in the universities and polytechnics (McLoughlin, 1973).

All of this had been achieved by the middle of the 1960s and the "night of the long knives" was ample confirmation of the power of the professional ideology. All that was lacking was a theoretical position on which to base the claim for a unique skill. This was provided at the turn of the decade by the various major texts which constitute the "systems approach" with their emphasis on the planning process itself, on operational research, quantified techniques and a claim to value-free objective analysis. These claims put their proponents in a position of considerable power both inside and outside the public service.

Such power in the bureaucracy has been bought at a considerable price. Large reservoirs of talent have been shut out or kept down by the insistence of employers on a professional qualification. Educational resources have been misused by the provision of courses (especially graduate courses) for students with admirable and relevant first degrees but whose main motivation to obtain the award was based largely on the need to obtain a professional qualification. The ideological necessity to claim a unique skill fostered the emphasis on abstract procedural theories and a severe neglect of substantive issues in urban and regional problems and policies. The need to foster the notion of a distinctive skill led to excessive time spent on "practical" work which was usually highly impractical but so time-consuming as to leave little room or motivation for research. (8)

British planners have visibly failed to deliver the goods. They have had little to say about the effects of world economic forces on our cities and regions, have poorly understood the workings of the property market, the housing market or more generally the ways in which public policies allocate life chances in differential ways. Planning educators have failed to do much research on such issues, whilst the research which has done so was carried out elsewhere, remote from the planning students themselves. Above all, planners in practice and in academic life have failed to study the place of land ownership, land tenure, land development and redevelopment in a market economy, and the limits placed upon public sector intervention and the role of the state generally. Yet knowledge of such substantive matters could have supported true claims to skill and expertise and a right to be heard in the corridors of bureaucracy.

WHERE TO NOW?

Morale amongst British planners and academics is at a low ebb. They seem unwanted and unloved by the general public, by conservatives and radicals alike and certainly by the present government. It is sometimes tempting to regard town planning as fit for another razor - not of the Margaret Thatcher or Malcolm Fraser kind but the far older and more respectable one of Thomas Occam - to see planning as a "gratuitiously multiplied entity". This would be a more serious item for discussion if governments were to abandon physical planning altogether. But there is no real prospect of that. Both capitalist and socialist countries, in all their diversity, find too much value in the regulation of land use and development. In every kind of polity and economy it is possible to have wretched, mean, unsocial, graceless arrangements of housing, shops, school and transport and also possible to have pleasant, gracious, sociable and human ones (Stretton, 1979).

The basis for achieving human living and working environments is a sensitive and carefully conceived interaction between the state and the market. This requires close links between practice, research and education. Research should in future emphasise a number of important themes or fields of enquiry. These include:

- the place of land in economic change and capital accumulation;
- the land development and conversion process;
- the economic and spatial dynamics of urban and regional change;
- the geographical and social class consequences of various kinds of change;
- the effects of different kinds of public policies on the social and environmental fabric of towns and regions;
- the relative effectiveness of different modes of state intervention and various kinds of planning policies.

Much more research of this kind should be carried out in planning schools. This is not to say economists or sociologists should do less - quite the reverse. However, if planning students are to learn more about the political, economic and social realities of urban and regional development they should be taught by people who are themselves actively and regularly engaged in critical and scholarly studies of practice. What this does not require is the stultification which too often results from misconceived notions of professionalism. The continuing dialogue between theory and practice which is necessary to achieve this requires firm academic leadership committed to a critical and scholarly tradition. It needs money, young enthusiastic teachers and researchers, and above all, good students who will develop in a vital and enquiring atmosphere.

ACKNOWLEDGEMENTS

I am most grateful to the (British) Social Science Research Council (SSRC) for commissioning the review on which this paper is based and to Andrew Broadbent, Lyn Davies and Patsy Healey for preparing background papers on research, practice and education (respectively) as inputs to the review. Neither they nor the Social Science Research Council are in any way responsible for the opinions I express which are entirely my own responsibility. These papers will eventually be published by the Social Science Research Council, I, Temple Avenue, London EC4Y OBD.

NOTES

1. A standard, but somewhat uncritical, account is to be found in Cullingworth (1977).
2. The flavour can be discerned by a perusal of McLoughlin (1969), Chadwick (1970), and Friend and Jessop (1969).
3. See Report of the Committee on the Qualifications of Planners, (1950). This remarkable historical document is a lucid analysis which remains after more than thirty years as fresh as the day it was written.
4. For a comment on the full "flowering" of the Rothschild doctrine under the Thatcher administration see the second editorial in Times Higher Education Supplement, Oct. 9, 1981.
5. For a critical but somewhat partial, account of planning research in the mid 1970s see Eversley and Moody (1976).
6. The basis for most of these remarks is the background paper by Andrew Broadbent prepared as part of the Social Science Research Council review, see note (1) above.
7. See note (6) above.
8. See Reade (1981) for a detailed analysis of the ideological function of "practical work".

REFERENCES

Chadwick, G. (1970) A Systems View of Planning. Oxford, Pergamon.
Cherry, G. (1974) The Evolution of British Town Planning. Leighton Buzzard, Leonard Hill.

Cullingworth, J. (1977) Town and Country Planning in Britain. 6th edition, London, Allen and Unwin.
Drake, J., McLoughlin, B., Thompson and Thornley, J. (1975), Aspects of Structure Planning in Britain. Research Paper. No. 20. London, Centre for Environmental Studies.
Eversley, D. and Moody, E. (1976) The growth of planning research since the early 1960s. London, Social Science Research Council.
Friend, J. and Jessop, W. (1969) Local Government and Strategic Choice. London, Tavistock.
Hall, P., Thomas, R., Gracey, H., and Drewett, R. (1973) The Containment of Urban England. London, Allen and Unwin.
Harris, B. (1966) 'Some problems in the theory of intra-urban location', Operations Research, 9:695-721.
Harris, B. (1966) 'The uses of theory in the simulation of urban phenomena', Journal American Institute of Planners, 32:258-73.
Harris, B. (1966) 'The content of a planning theory course', Oxford Polytechnic, Department of Town Planning, Working Papers in Planning Education and Research. No. 25.
Kantorowich, R. (1967) 'Education for Planning', Journal of the Town Planning Institute, 53:175-181.
Llewelyn-Davies, R. (1967) 'Research for Planning', Journal of the Town Planning Institute, 53:221-25.
McLoughlin, J.B. (1969) Urban and Regional Planning: a systems approach. London, Faber and Faber.
McLoughlin, J.B. (1973) Control and Urban Planning. London, Faber and Faber.
McLoughlin, J.B. (1973) 'the future of the planning profession', in Cowan, P., ed., The Future of Planning. London, Heinemann.
Ministry of Housing and Local Government (1965) The Future of Development Plans. London, HMSO.
Reade, E. (1981) 'Practical work in town planning education', Oxford Polytechnic, Department of Town Planning, Working Papers in Planning Education and Research. No. 54.
Stretton, H. (1978) Urban Planning in Rich and Poor Countries. Oxford, Oxford University Press.

3 On the shoulders of which giant? The case for Weberian political analysis

Peter Saunders

It is notable that, of all the social sciences, sociology appears most obsessed with the work of its 'founding fathers'. Economists read Smith and Ricardo out of historical curiosity; political scientists do not generally become embroiled in fierce debates about what Hobbes or Mill 'really' said; but sociologists every year devote many hours of lectures and many pages of print to intensive discussions of the work of Marx and Weber.

Doubtless this peculiar feature of modern sociology is due as much to the immaturity of the discipline as to the continuing relevance of Capital and Economy and Society for an understanding of contemporary problems. Certainly Payne and his co-authors in a recent diagnosis of the ills of the discipline in Britain have expressed their despair at what they see as "an almost neurotic desire to legitimate current endeavours by locating them in a long and glorious tradition and at worst, an attempt to disguise present poverty by hiding in the apparent richness of past contributions" (Payne, Dingwell, Payne, Carter, 1981:41). These authors go on to argue that this preoccupation with the theories of nearly a century ago has reached pathological proportions and goes a long way in explaining the weakness of sociology today; sociology, they suggest, must evolve into a more policy-relevant and empirically-based discipline if it is to survive and to realise its promise.

I have considerable sympathy for this view, yet, parodoxically perhaps, I wish to devote this paper to a discussion of marxist and weberian approaches within the field of urban political analysis, and in particular to an examination of the basic principles of a weberian perspective. I justify this indulgence on three grounds. First, the theories and methodologies associated with Marx and Weber do help to define the two basic and competing approaches within urban studies, yet while Marx's work has received considerable attention in the recent literature, Weber's has rarely been discussed at all. Secondly, weberian sociology has, as a consequence of this, been widely misinterpreted and even caricatured in recent marxist critiques. Thirdly, there is evidence of a resurgence of a broadly weberian approach to urban politics which, unlike some of the marxist work in this field, does have the potential for generating empirical work with direct relevance for contemporary issues. Some of the weaknesses of this work (notably regarding its treatment of 'corporatism') can, I believe, be clarified and overcome through a firmer understanding of Weber's political sociology.

A new orthodoxy?

For the last ten years, the most interesting and productive developments within urban studies have arguably issued from various versions of marxist theory (notably, of course, the work of Castells and Harvey). While such work has not displaced or overthrown more traditional approaches, it has seized the initiative in theoretical and methodological debate and has become the dominant perspective to which other innovatory work has had to respond. It need hardly be added that this rise to prominence of critical materialist analyses has been facilitated by the decline into crisis of most of the world's major capitalist economies, a tendency which has rendered problematic the old orthodoxies of neo-classical economies, liberal democratic political theory and functionalist sociology. It generated a crisis within non-marxist social science while at the same time regenerating interest in marxist theory.

The time has come, however, for a critical reflection on the new marxist orthodoxy. I say this for two reasons. First, there has been too ready a tendency to draw simple and misleading polarities between marxist and 'bourgeois' perspectives in urban studies. While various writers have identified important divergences within the marxist literature (see for example, Pickvance, 1977, Harloe, 1979), the differences between alternative non-marxist approaches have too often been disregarded, with the result that competing perspectives have been subsumed under single all-encompassing and usually negatively-charged labels. In this way it has been possible to dismiss all alternatives to marxism through a critique of any one of them.

Weberian urban sociology has been one notable victim of this process, for methodological attacks on idealism and theoretical attacks on pluralism have been employed to dismiss weberian approaches when Weber's own work was, I would argue, neither idealist nor pluralist. The time has come, therefore, to stand back and re-examine the credentials of a weberian approach seriously, and to stop setting Weber up as a straw man to be knocked down by the allegedly more subtle claims of a materialsit analysis. The need for such a reassessment is made all the more pressing by the appearance of a growing number of books in this field which identify serious problems in contemporary marxist analysis and which attempt to develop an alternative approach (see for example Dunleavy, 1980, Goldsmith, 1980, Simmie, 1981 and my own recent work, especially Saunders, 1980 and 1981a).

The second reason for a critical re-evaluation of marxist urban theories at the present time is that such work has, I believe, come to the end of a cul-de-sac from which, given its core premisses, it cannot escape. Much has been learned from the journey over the last ten years or so, but the destination has eluded us. I shall argue in this paper that ultimately the methodological foundations of the various versions of marxist urban theory are by no means as secure as many writers have asserted, and that the fundamental principle of some notion of economic determinacy (which lies at the heart of any marxist perspective on politics) must effectively preclude a solution to the problems of contemporary urban analysis which this work has itself identified.

The paper therefore begins with a critical discussion of methodological and theoretical problems in the contemporary marxist literature, proceeds to an elucidation of weberian methodology and political sociology, and concludes by examining some recent attempts to apply a broadly weberian sociology to the analysis of urban politics today. I shall not here be developing my own framework for urban political analysis, for this is available elsewhere, although I do see this paper as, in part, an elaboration of the principles which are embodied in that framework.

The limitations of contemporary marxist urban analysis

A critical evaluation of the current state of marxist urban analysis is a hazardous undertaking if only because there are so many varieties of contemporary marxist discourse (see Lebas, 1982 for a review of some of this work). The work which has undoubtedly had the greatest influence in urban studies over the last ten years has tended to be that which derived from the writings of the French marxist philosopher Louis Althusser in the 1960s - Castell's original critique of 'bourgeois' urban sociology, for example, was to a large extent grounded in Althusserian epistemology (in particular the science/ideology distinction), while analyses of state intervention in the urban system have often owed much to the work of Poulantzas whose early work was premissed on the Althusserian conceptualisation of modes of production. However, Althusserian marxism has itself been attacked, both in terms of its methodology and its theoretical foundations. Castells and Poulantzas have both attracted widespread sympathetic criticism in marxist circles and have both in their later work exhibited a growing distance from their earlier formulations. Meanwhile, other writers have developed other broadly marxist approaches by emphasising other aspects of Marx's work (for example, David Harvey's emphasis on the circulation of capital, Henri Lefebvre's concern with the quality of life, recent work in America by Alford, Friedland and others on fiscal crises and problems of legitimation, and so on).

Clearly no one writer and no one body of marxist literature can be said to represent 'the' marxist approach to urban analysis. Yet having said that, I would suggest that all marxist approaches share in common to a greater or lesser degree a certain methodological and theoretical commitment which has become problematic. If we were, in a sense, to construct an ideal type of marxist theory, then in terms of its methodology it would make some claim to privileged knowledge vis-a-vis non-marxist approaches, and in terms of its theory it would be committed in some way to the principle of economic determinancy. Different writers argue these principles in different ways, but it is my contention that what is distinctive to marxist writing in general is its commitment to the notion of its own superior knowledge and to the primacy of the economy in social affairs. The problem, quite simply, is that these two basic elements of marxist discourse can no longer be sustained. The claim to epistemological privilege is groundless while the commitment to economic determinancy generates an irresolvable problem of theorising the autonomy of politics.

Problems of method

Marx's method of dialetical materialism entails two fundamental axioms. The first is the principle of the dialectical unity of opposites, from which is derived the emphasis on analysing society as a totality. No one aspect of social life can be understood except in relation to all other aspects with which it forms a contradictory unity, and it follows from this that any analysis must be premissed on a priori theory of how the totality is structured.

The second axiom is that the material world exists prior to and irrespective of our ideas about its existence. Furthermore, while our consciousness of the world must bear some relation to reality, this relation is often distorted, obscured or even inverted. It is precisely that task of science to penetrate the 'phenomenal forms' in order to identify the 'essential relations' through which material reality is structured and reproduced: "The forms of appearance are reproduced as current and usual modes of thought: the essential relation must first be discovered by science" (Marx 1976:682).

It follows from these two axioms that a claim to knowledge will be ideological rather than scientific to the extent that it either presents a partial picture of reality unrelated to the whole, or reproduces our ideas about reality without discovering the essential relations which lie behind these ideas. A science, in short, must understand the inter-relations which structure the world and the underlying mechanisms that generate them, and the so-called 'bourgeois' social science disciplines fall short on both criteria. Economics, for example, fails to grasp the totality of social (class) relations within which processes of production, distribution and exchange are embedded (hence Marx's derisive comments on the Robinson Crusoe assumptions held by classical economic theory), and its emphasis on the phenomenal categories of market exchange prevents any deeper understanding of the operation of the law of value which underpins such exchange.

All this raises two obvious problems. First, how is marxism able to grasp the complexity of social processes in their totality when other approaches only ever achieve a partial explanation? Second, how is marxism able to penetrate to the essential reality when other approaches remain stuck at the level of appearances? There are basically two types of response from within the marxist literature when faced with these questions.

The first derives from Althusser's reading of Marx and was carried over into urban studies in Castell's earlier work. According to Althusser (1969), debates in epistemology have revolved around a false duality between empiricism (the argument that knowledge is a product of pure experience) and idealism (the argument that knowledge is a product of human consciousness and thought), a duality which Marx overcame through what Althusser terms 'theoretical practice'. It is the task of theoretical practice to transform ideological raw materials of thought into scientific knowledge, and this is achieved through a process of production of knowledge in which the individual scientist applies a set of conceptual tools to the raw material (existing ideological representations of the world), just as an individual worker may apply a set of tools to a raw material in order to produce a physical commodity. What is important to note about this process is that, just as subjectivity of the worker on the production line is irrelevant to the determination of the end product since he or she is merely an agent of the means of production applied, so too the subjectivity of the scientist is of no consequence for the end product which is merely the result of the application of a determinate mode of analysis to existing raw materials of thought.

It was this rather curious epistemology which Castells brought to bear in his early critique of urban sociology as 'ideological' and in his subsequent reformulation of the urban question in which he reworked existing conceptions of space and collective consumption from within a marxist framework. It is, however, an epistemology which rests on the mere assertion that Marx's method (as interpreted by Althusser) is in some mysterious way 'correct'; that the application of marxist concepts to existing ideas will produce scientific knowledge while the application of non-marxist concepts will serve only to reproduce ideology. It is, furthermore, a recipe for smug armchair theorising, for if we cannot know the world directly (the empiricist fallacy), then empirical research loses all point and scientific work can only mean picking over existing 'bourgeois' theories and subjecting them to a materialist critique. Little wonder, then, that Payne and his colleagues despair at the influence of Althusserian marxism in British sociology, for it is not simply non-empirical but is anti-empirical.

Castells, of course, has in his later work largely abandoned the Althusserian epistemology with which he launched his initial critique of previous urban sociology. Following his experience of conducting empirical research in the Dunkirk region of France, the split with Althusser became ever more pronounced. By 1975, when he wrote the Afterword to the English edition of <u>The Urban Question</u>, he was able to state that, "Theory is not produced outside a process of concrete knowledge" (1977:438), and this was followed later by an auto-critique in which he criticised himself and others for applying "established theories without modifying them to the reality observed" (1978:11).

Today, therefore, Castells seems to have arrived at a position where he seeks to develop theory through empirical observation and vice-versa, a position totally divorced from Althusserian epistemology. The problem now is that, having abandoned the earlier commitment to Althusser, he has also lost the epistemological foundation of his original critique of alternative approaches within urban sociology as 'ideological'. In other words, we are drawn to the conclusion first, that Castell's earlier epistemology could not justify its own claims to priviliged knowledge (since these rested on mere assertion), and secondly, that in his subsequent abandonment of this epistemology, Castells has in any case undermined the already shaky foundations of his claims to have founded a uniquely scientific urban theory.

A second and potentially more fruitful response to the question of how marxism can lay claim to peculiar insight into the inner workings of the social system derives from what has come to be known as the 'realist' philosophy of science as discussed by Bhaskar (1975) and others.

For Bhaskar, the task of any science must be to discover the necessary tendencies inherent in things, irrespective of whether these tendencies become manifest in particular situations where various contingent factors may be operating. It is inherent to the molecular properties of copper, for example, that it conducts electricity, and this scientific proposition cannot be falsified by empirical cases where copper fails to conduct electricity current due to contingent factors (for example, a dirty connection). Similarly, to take an example cited by Andrew Sayer (1979), the law of gravity is not disproved by the observation that aeroplanes do not usually fall out of the skies.

According to Bhaskar, science discovers the inherent generative properties of things through a three-stage process. First, it involves observation of recurring patterns of events in the world (for example, apples tend to fall downwards; capitalist economies tend to go through successive periods of crisis). Second, it involves theoretical speculation on the inherent tendencies which may be generating the phenomena observed (for example, apples are subject to gravitational force, capitalist economies are subject to a tendency for the rate of profit to fall as the organic composition of capital rises). Third, such theories are subjected to experimental evaluation involving the removal of all contingent factors under laboratory conditions in order to discover whether the posited tendency or mechanism does in fact explain the phenomena observed.

It is at this third stage, of course, that natural science and social science (including marxism) part company. As Bhaskar himself recognises, "The central argument...has turned on the possibility of experimental activity...it is clear that experimental activity is impossible in the social sciences...there is a general problem of confirmation (or corroboration) and falsification in the non-experimental sciences" (1975:244-5). Yet those marxists who have sought to establish the scientific character of their insights by reference to realist philosophy have failed to appreciate the crucial role played by experimentation in this philospohy. Effectively, what we have in marxism is the claim to be able to identify underlying and inherent mechanisms in society yet the refusal to accept any possibility of empirical tests on the grounds that contingent factors may be operating. We are told, for example, that there is an inherent tendency for the rate of profit to fall as capitalism develops but that the operation of counteracting tendencies denies any possibility of empirical confirmation or refutation of this 'law'. As Andrew Sayer puts it, "In social science there are no tests, only applications" (1979:33).

Such reasoning produces no more than self-confirming tautologies (for example, if profit rates are falling, this is due to the inherent tendency for the rate of profit to fall, and if they are not, this is due to the effect of counteracting tendencies). One is reminded of the story of the man on a train journey through the Home Counties who scattered mustard seed from the window in order to keep the elephants away. When told there were no wild elephants in Surrey, he replied that this only served to demonstrate the efficacy of mustard seed as an elephant deterrent.

The most that a marxist method can achieve (and it is extremely fruitful and innovative so long as its limitations are recognised) is conjecture - that is, positing the existence of certain tendencies and generative mechanisms which, if they did exist, would help to explain phenomena observed. This is what Derek Sayer has termed a 'retroductive method': "The logic of Marx's analysis is essentially a logic of hypothesis formation, for what he basically does is to posit mechanisms and conditions which would, if they existed, respectively explain how and why the phenomena we observe come to assume the forms they do" (1979:114). But such a method can never demonstrate that a posited tendency does actually exist and can sustain no claim to epistemological privilege since other theorists may equally posit other possible explanations for observed regularities.

This being the case, there is clearly no warrant in marxism for claiming its own method as 'scientific' and dismissing all others as 'ideological'. Indeed, there are no grounds outside of marxism itself for insisting that theories must involve conceptualisation of the totality, or must attempt to penetrate appearances in order to discover essential relations. To dismiss other theories as ideological through the application of criteria of scientificality which are internal to marxist discourse is clearly fallacious, for there is no epistemology which can be considered binding outside of specific discourses and which can legislate on which knowledge is to be considered 'scientific' and which 'ideological'. As Hirst argues, "marxism is not a science (equally it is not a non-science; science-ideology is an epistemological distinction), it has no privileged knowledge" (1979:6). It follows from this that different perspectives in the social sciences must be evaluated on their own terms. It is no longer good enough to attack alternative approaches because they are 'idealist', 'positivist' or whatever, for there is no epistemology which can be set up as the yardstick against which all approaches must be measured for scientific adequacy. The social sciences are, and must remain, pluralistic.

Problems of theory

The logic of my argument so far is simply to establish the 'right to coexistence' of perspectives other than marxism within urban studies. My argument is, however, stronger than this since I believe that, judged on its own terms, marxist theory cannot resolve the key problem which lies at its heart, namely the economics/politics relation. This problem was succinctly summarised by Poulantzas, one of the most influential marxist political theorists of recent years, in an interview given shortly before his death, where he admitted that, "The problem still remains, how to find the specificity and the autonomy without falling into the absolute autonomy of politics. It is the core of the marxist problematic" (*Marxism Today*, 1979:196). It is my contention that, given the centrality of the principle of some form of economic determinacy in marxist political analysis, this problem not only "still remains" but cannot be resolved.

There are three main strands of thought in contemporary marxist political analysis which attempt to overcome the problem. These are:

i. Instrumentalism/economism - the argument that politics are directly reducible to economics, that the state is thus little more than an instrument of the dominant economic class and that political struggles can be analysed in economic class terms (for example, as struggles between capital and labour).

ii. State derivationism - the argument that, in capitalist societies, politics and economics have evolved as separate spheres of practice, this reflecting the needs of capital for a formally autonomous and supposedly neutral state form through which class relations can be regulated in an authoritative manner.

iii. Relative autonomy theories - the argument that politics are not directly reducible to economics, that political struggles 'represent' classes and that the state must function in the long term interests of the dominant fraction of capital while in the short term responding to the demands of non-capitalist classes.

These theories have been chewed over by many writers (myself included) on many occasions, and I shall not go into detail here. The main point is simply that all three collapse sooner or later into an unjustifiable economistic interpretation of politics and thus fail to explain the autonomy of the political sphere.

The problem is clearest in instrumentalist theories, for if you believe that the state is "nothing but a machine for the oppression of one class by another" (Engels), or that it is "an organisation of violence for the purpose of holding down some class" (Lenin), or, to take a more recent formulation, that "the state should be considered as the instrument of the ruling class" (Miliband), then it becomes exceedingly difficult to explain those situations in which capital has apparently lost while other classes have won as a result of state policies.

Instrumentalist theories generally respond to this problem by suggesting that apparent working class victories are at worst 'actually' defeats, or are at best concessions won without cost for the bourgeoisie (see, for example, the discussion of the history of British housing policy this century by Dickens, 1977). Such a view usually rests on some assumption of ruling class canniness and foresight, the state buys off future trouble with a few insignificant trifles - a National Health Service here, free universal education there, low cost public housing for a third of the population... The problems with such an analysis are obvious, for not only are the 'concessions' far from trifling, but the existence of a far-sighted and politically-sophisticated ruling class seems (in Britain at any rate) somewhat implausible. Evidently, then, the state cannot be considered as a mere pawn in the hands of the bourgeoisie.

State derivationist theory agrees. Hirsch (1978), for example, recognises that, unlike feudal society where the institutions of law and politics are directly tied to and controlled by the land-owning class, the state in capitalist societies enjoys a formal autonomy from the capitalist class. However, he argues that the very form of capitalist politics and law is itself inherently bourgeois, for the capitalist class needs an autonomous political sphere standing apart from the labour-capital relation, in order to reproduce that relation. The essence of capitalism is the sale and purchase of formally free labour-power, thus a legal system is required which constitutes workers as individual legal subjects rather than as, say, bonded serfs. Similarly, capital needs a political system grounded in the illusion of individual citizenship in order to fragment a potentially threatening working class into millions of atomised voters.

The problem with this elegant approach is that it falls foul of the functionalist fallacy of transposing cause and effect. Even if it can be demonstrated that capital does in fact 'need' certain political and legal forms in order successfully to reproduce itself, this can in no way explain how such needs came to be met in the first place. Ultimately we are driven back to familiar economic reduction; politics take the form they do because they reflect the economic requirements of the capitalist class. Far from resolving the problem of the economics/politics relation, this approach reaffirms it with a vengeance - political forms reveal the logic of capital, and the theoretical categories of political analysis are directly deducible from the theoretical categories of economic analysis.

Relative autonomy theories (notably, of course, in the work of

Poulantzas and derivatively of Castells) do recognise that the state cannot be explained purely in functional terms as serving the needs of some class and that political struggles cannot necessarily be analysed directly in terms of the capital-labour relation. The problem with these theories, however, is that while they recognise the relative autonomy of politics, they fail to explain it.

Part of the problem is that the very notion of a 'relative autonomy' of politics lends itself to easy tautology - if the working class wins, this can be explained in terms of the relative autonomy of the state from capital; if the bourgeoisie wins, this can be explained by the fact that this autonomy is only relative! There is, one suspects, little that the state might do which could be used as a check on the validity of such a theory.

The problem, however, goes deeper than this and concerns the attempt to combine a structuralist and determinist theory of politics with a recognition of the significance of class struggle as an agency of political change. Wearing his structuralist hat, Poulantzas insists that the role of the capitalist state must be to maintain the interests of the dominant class in the long run, even at the expense of short term concessions. Wearing his activist hat, however, he also insists that the state is a condensation of political class struggles and that its policies at any one time will reflect the balance of power between the classes. The problem which he fails to address is how the varying fortunes of class struggle inevitably result in state policies which are consistent with the interests of the dominant fraction of capital.

The assumption on which his argument seems to rest is that working class mobilisation will bring about a reformist response on the part of the state which will in turn dissipate further struggle and thus pave the way for a reassertion of capitalist hegemony. This is not an unreasonable hypothesis (the period 1974-8 in Britain would seem to 'fit' quite well), but history suggests that events need not always follow such a sequence (successful working class action may rather revolutionise aspirations and lead, not to dissipation, but to more and radical demands). Clearly there is no necessity in Poulantzas's theory of the relation between the causes and functions of state policy-making.

A further problem concerns the relation between political struggles and class struggles. Poulantzas recognises that political groupings such as parties and pressure groups cannot be seen simply as the embodiment of economic classes, yet he seeks to maintain a class analysis of political movements through the fudgy concept of 'representation'. Classes, he suggests, are represented in politics even though classes are not directly reducible to the economic categories of capital and labour. This is entirely unsatisfactory as a solution to the problem, for the concept of representation becomes useless when we turn to empirical analysis of actual political movements. In what sense do the various Loyalist groups in Northern Ireland represent a class, and which class do they represent? And what of those groups with which urban sociology has been particularly concerned, groups such as tenants' associations, public transport users, students and other consumers of state services? Either these are or they are not class movements, and to say that they 'represent' classes in some vague and unspecified way is to say nothing. Clearly what is required is a theory of politics which can distinguish those areas in which class is a valid concept from those in which it is not.

What this brief review of three different approaches within marxist political theory has shown is that marxism has failed to resolve the problem of how to reconcile the fact of (at least partial) political autonomy with the core principle in all marxist discourses of economic determinacy. Relative autonomy theories have perhaps come the closest to an answer, but they too have eventually collapsed into economism. As Hirst (1977) has shown, the problem of the relation between economics and politics is a problem which allows of only two solutions - either economism or what he terms 'necessary non-correspondence' - and faced with this choice, marxist theories have inevitably relapsed into the former since acceptance of the latter involves abandoning the principle of economic determinacy which in one form or another is an essential hallmark of marxist analysis.

An economistic theory of politics is, I would argue, untenable, for capitalist economies have sustained many different types of political systems, capitalist governments have pursued many different types of policies and political struggles have consistently been waged over many different types of issues, many of which cannot be understood with reference to the struggle between wage labour and capital. If we reject the principle of economism, then we are left with the principle of non-correspondence - that is, an historically contingent relation between economics and politics in which the one cannot be deduced from the other. Acceptance of this principle points us away from Marx and towards the political sociology of Max Weber.

The potential of weberian sociology

Weber has suffered more than most from misrepresentation and misunderstanding in recent debates within urban studies. In particular, his methodology has been dismissed as 'idealist' and his political sociology has been attacked as 'pluralist'. My objection to the first of these apparently derogatory labels is that it is both misguided and irrelevant; misguided because Weber's use of ideal types has been confused with idealist epistemologies, and irrelevant because, as we saw above, it is not possible to demonstrate the inadequacy of a given discourse through a critique of its epistemological foundations. My objection to the second line of criticism is simply that it is wrong, for any careful reading of Weber's political sociology will reveal that there is no notion of the neutrality of politics such as characterises pluralist analysis. I shall briefly consider each of these points.

Weber's method: ideal types are not idealism

To the extent that marxist urban theory has addressed Weber's work, it has tended to dismiss it as hopelessly idealist - that is it has been argued that for Weber, knowledge is a product of our ideas about the world, in which case weberian sociology fails to address itself to an analysis of the material reality of which such ideas are merely an expression. A good example of this line of argument is found in Hindess's caustic comment that for Weber, "There is no reason why the social scientist should not let his imagination run wild. He has nothing to lose but the chains of reason" (1977:38). Similar arguments have been developed in the context of urban studies by Andrew Sayer (1980) and Hooper (1982), both of whom believe that by making a case against idealism, they have succeeded in establishing an argument against weberian ideal type analysis.

This is curious, for if we must hang labels around people's necks, I would describe the main thrust of Weber's methodology as 'empiricist' in the sense that he believes that knowledge of the world is generated through observation. The problem which he identifies is not that reality cannot be known, but is rather that, in observing reality, we are confronted with a chaos of sense impressions which have to be sifted and ordered by means of concepts before a sociological account can be attempted. The role of ideal types, therefore, is not to generate knowledge but to make knowledge possible by means of prior conceptualisation which enables us to classify phenomena and to sort out that which is relevant for our purposes from that which is not.

Any ideal type will inevitably be partial, for in its construction we exaggerate and logically purify those aspects of reality which we are interested in studying and disregard the rest. It is in this sense that sociology is grounded in the values which give research its coherence and direction: "There is no absolutely 'objective' scientific analysis of culture or viewpoints according to which - expressly or tacitly, consciously or unconsciously - they are selected, analysed and organised for expository purposes" (Weber, 1949:72).

This insistence on the inevitable partiality of any sociological account and on the impossibility of knowing the totality has naturally attracted criticism from marxist writers. Mingione, for example, has recently suggested that, "Typologies may seem convenient for a better understanding of social realities but they do not take into account the complex connections between different processes and situations and miss their causal relations ... Historicist typologies, no matter how articulated and precise they are, cannot explain the causal origins of social phenomena" (1981:71). Yet this misses the point of Weber's methodology in two crucial respects.

First, Weber argues that no approach can hope to take into account all the "complex connections between different process" in society. Any one social phenomenon has an infinite multitude of historical causes, some of which are more significant than others, but all of which have contributed to making that phenomenon what it is. It is hardly sensible, then, to criticise Weber's methodology for its failure to be all-encompassing when it rests on the argument that theories of social totality cannot be constructed, and when we have seen in the previous section that various marxist attempts to relate political and economic phenomena through a general theory have failed.

Second, Mingione fails to understand that weberian typologies are not intended to "explain the causal origins of phenomena". The construction of ideal types is simply the indispensible first step towards explanation. Having constructed them, and thus having rendered the complexity of the world intelligible, we must then go on to develop hypotheses which have to be tested and evaluated in terms of historical and comparative research aimed at assessing their causal adequacy. An ideal type itself tells us nothing we did not already know, but it is impossible to develop our knowledge without it.

The application of a weberian methodology to urban sociology will thus involve first, an ideal typical conceptualisation of those aspects of the social world in which we are interested and which may be identified (purely in a nominalistic sense) as 'urban', and second, the development of a theoretical argument which attempts to explain why these phenomena take the form they do at particular times in particular places. Thus in my own recent work, I have attempted both to develop an analytical framework for urban political analysis (Saunders, 1980 and 1981a), and to apply this framework in order to develop a theoretical understanding of issues such as conflicts between central and local government in Britain (Saunders, 1982) and the sacking of the Melbourne City Council in Australia in 1981 (Saunders, 1981b). One aim of the present paper is to clarify and (in the face of continuing criticism) to justify the methodological principles which underpin such work.

Weber's political sociology: autonomy is not neutrality

Just as Weber's method has been misrepresented in recent debates, so too has his political theory. Part of the reason for this may be that critics have rarely addressed themselves to the primary sources, but have more often attacked 'weberian sociology' through a critique of contemporary writers (notably Pahl) whose work is seen to exemplify the weberian tradition. However, Pahl's work too has sometimes been glossed over or misinterpreted, with the result that there has been a tendency to equate weberian political sociology with pluralist theory and thus to extend to the former criticisms which have in the past been levelled against the latter.

The key elements of liberal-democratic pluralist political theory are as follows:

i. an emphasis on the essential neutrality of the state, as in Dahl's market analogy in which the state is likened to a political market place in which the supply of policies comes to reflect the pattern of expressed effective demand from different consumers;

ii. an assumption of a fundamental value consensus in society, as in Dahl's assertion that, "Prior to politics, beneath it, enveloping it, restricting it, conditioning it, is the underlying consensus on policy that usually exists in the society among a predominant proportion of the politically active members" (1956:132);

iii. a belief in the crucial significance of elections as the means whereby popular will becomes translated into state policies and those in power are held accountable to the masses, as in Birch's argument that elections ensure that politicians "respect public opinion and adjust their policies to intimations of public need" (1964:236).

None of these three principles are found in Weber's political sociology, nor in Pahl's writings on urban managerialism and corporatism, yet marxist critics have suggested (in relation to Pahl in particular) that his work "begs, it would seem, all the questions of the pluralist model of political power of which it is a variant" (Lambert, Paris and Blackaby, 1978:11), that, like the pluralists, he believes that state power "derives from the consent of the governed" (Hill, 1977:42), and so on.

It is not my intention here to defend Pahl's work, for as we shall see, I believe there are real problems in his methodology and his theory which serve to weaken his analysis of urban politics. Nevertheless, it is important that in criticising his work, we do not caricature it. It is also important that we do not reject weberian political sociology on grounds which may be appropriate to Pahl's particular application of some of Weber's ideas but which do not apply to Weber himself. Just as marxists would be justifiably aggrieved were we to dismiss the whole of Marx by means of an attack on Miliband, so too we should take care not to dismiss Weber through criticisms of one or two later writers who have attempted to apply some of his insights in a particular way.

Now it is true that unlike Marx, Weber argues for the analytical autonomy of economics and politics (which is why I see Hirst's argument for a 'necessary non-correspondence' between the economic and the political as more weberian than marxist in character). Such an argument does not deny the often close interrelationship between these two aspects of social life, for Weber is well aware throughout his historical works of the important influence of different types of political-legal systems on economic development, and vice-versa. What Weber does insist on, however, is that the relationship between economic and political forms has to be established at different times and places through empirical research, and cannot be deduced from any *a priori* deterministic general theory. Similarly, he accepts that those with economic power have often also been politically dominant, but again he insists that there is no necessary reason why this should always be the case.

To understand his argument, we need to begin with his basic

axiom that the unit of sociological analysis is the acting subject in relationship with others and not (as in marxist discourse) social entities such as classes. For Weber, collective concepts such as 'class' or 'state' are in principle always reducible to aggregates of individual subjects pursuing their goals and values. It is not classes or states which act, but groups of individuals who together may be termed a class, a political leadership stratum, or whatever. Sociological explanation must therefore always begin with an analysis of the goals and values which comprise individuals' motives for action.

It is a fundamental tenet of Weber's entire sociology that values - the realm of moral choices and ethical dilemmas - are always contestible (hence, of course, the emphasis in his methodology on separating questions of fact from issues of evaluation, for science can never resolve moral questions). It follows that, wherever different actors come into contact with one another, there will be a likelihood that they will be pursuing different and incompatible values. Once one actor attempts to realise values which others do not share, conflict becomes inevitable. This is why the concepts of 'power' and 'domination' are so central to Weber's sociology, for power determines whose values will prevail and whose will be ignored. When the ability of one actor to achieve his or her will becomes regularised, such that there is a typical probability that he or she will always prevail in a particular situation, Weber talks of a relationship of domination.

The ability to dominate others will in general rest on one of two bases; either control of market opportunities (that is, economic or class power) or control or influence over the state, the authoritative political association which alone has the right to enforce the compliance of others (that is, political power). These are analytically separate dimensions of power in modern society, which is precisely why Weber insists on the analytical autonomy of economics and politics.

If we now concentrate on the question of political domination, we see that for Weber, the state is an association controlled by particular individuals who pursue particular goals and values which are then imposed on others. Weber is not particularly concerned about the reasons why others obey. It may be out of fear, admiration, habit or self-interest since all that matters is the fact that they do, typically, obey. He recognises the importance of legitimation in relation to domination, but his concern is always with how those in power claim legitimacy rather than with whether and why those in subordinate positions cede it. There is nothing in Weber to suggest that those who control the state apparatus govern by consent, for given his emphasis on the inevitability of value conflict in society, genuine value consensus is clearly impossible. The point is, of course, that once an individual becomes an instrument of another's will, his or her values cease to be significant. He or she ceases, in a sociological sense, to be an acting subject, and becomes merely the receptacle of another's goals and values. All that matters for Weber is the fact of obedience. Thus, when he talks of 'consensus' he means simply the persistence of relations of domination (for example, "Persistent dissatisfaction endangers the stability of a coercive regime, but it does not invalidate the consensus as long as the powerholder can objectively count on the adequate execution of his commands" (1968:1378). Similarly, when he refers to the state as exercising a monopoly over the 'legitimate' use of force, he means simply that it alone can make a claim to such legitimacy, not that those subject to such force necessarily see it as legitimate.

It should be clear from all this that Weber's insistence on the autonomy of politics as a distinct sphere of domination in social life does not and cannot imply a thoery of the state as neutral. A political association with the right to use coercion cannot possibly be neutral if it functions as a means whereby one group of people can impose their goals and values on others who in all probability do not share them.

This conclusion is reaffirmed in Weber's discussion of modern representative democracy. He rejects the view that parliaments represent the 'will of the people', partly because direct democracy is impossible in complex modern societies, and partly because there can be no single will of the people. That Weber favours parliamentary forms of government is due to two largely pragmatic reasons. Firstly, in that the liberal democratic process helps to mobilise a groundswell of mass support for a particular political leadership - democracy and demagogy go together, for the votes of the masses are won through emotional appeals on the part of charismatic Caesarist leaders. Secondly, an effective parliamentary process facilitates the emergence of strong charismatic leaders who may be able to counter the inexorable yet deadening advance of bureaucratic domination over modern political administration. A healthy political system will strike some sort of balance between bureaucratic and charismatic domination, between the ethic of responsibility and the ethic of conviction, between value-rationality and instrumental-rationality, and a flourishing parliamentary system is the best hope for realising such a balance.

It is clear, then, that Weber stands poles apart from the defining principles of pluralist democratic theory with which he has so often been confused. While pluralist theory argues for the essential neutrality of the state, Weber sees in the modern state the means whereby one group may impose its values and interests. While pluralist theory sees the political process as grounded in a basic value consensus, Weber rejects the very possibility of such a consensus and sees the masses as the tools rather than the masters of the ruling group's will. And while pluralist theory sees elections as crucial in ensuring the accountability of the leaders to the led, Weber sees representative democracy as playing a limited role in manipulating consent and in countering the growing power of the appointed expert.

The state of weberian urban social theory

At the start of this paper, I noted that a number of books have been published in the last two or three years which have expressed dissatisfaction with the new marxist orthodoxy in urban studies and which have attempted to develop a fresh approach grounded to a greater or lesser extent in weberian sociology. In one of these books, Goldsmith has suggested that different perspectives on urban politics, including work in the marxist tradition, should be seen as complementary: "Each alone fails to provide a complete explanation of the city's political processes the possible virtues and contribution of other theoretical perspectives: the models should not be seen as competitive so much as complementary. We need to fuse our models together" (1980:46).

Such calls for theoretical pluralism (which should not be confused with eclecticism) are not fashionable in the social sciences today, but Goldsmith's argument is, I would suggest, entirely consistent with Weber's argument that we can only ever achieve partial and one-sided accounts of the social world. The hallmark of weberian sociology should be its modesty - a recognition of its own limitations and of the potential contribution of different perspectives to an understanding of different aspects of social reality. As he made clear at the end of his essay on the 'Protestant Ethic and the Spirit of Capitalism', it was never Weber's intention to substitute for a one-sided materialist (marxist) analysis a one-sided non-materialist one.

In urban studies today, however, there is a real danger of an over-reaction against the marxist orthodoxy involving a total rejection of marxist theory and the attempt to establish an alternative general and historical theory in its place. This tendency is perhaps manifested most markedly in a recent book by Simmie (1981), and it is rooted in the recent growth of academic interest in the question of corporatism. Corporatism, in other words, has been seized upon as the weberian answer to marxist political theories, rather than as one perspective among others, and as a result the concept has become over-generalised and almost meaningless.

Corporatist theory was introduced into urban studies primarily through the work of Pahl (see especially Pahl, 1977a, 1977b and 1979). The influence of Weber's political sociology was immediately apparent in Pahl's conceptualisation of the state as collectivity controlled by key individuals pursuing particular values (order, unity, nationalism and success), and in his emphasis on the role of technical experts in modern political administration. Pahl's work also reflected Weber's insistence on the analytical autonomy of politics and economics, but it went further than Weber ever did by also asserting the virtual independence of the political system from the economy in the conditions evolving in Britain in the 1970s. It will be recalled that Weber's insistence on the autonomy of economics and politics did not preclude a recognition of strong interrelations between the two - that is, although political processes could not be reduced to an economic determinism, Weber certainly recognised that economic factors could and did influence and limit political action. In his concept of a new corporate society, however, Pahl underemphasised the extent to which the state remains constrained by the capitalist economy, and the image which emerges from his work with Winkler (1974) is that of an independent political force in society directing economic action according to its own non-market principles.

Now it is important to distinguish the issues of autonomy, independence and neutrality in political analysis. Unlike pluralist theory, Pahl does not argue that the state is neutral, for he is well aware that consistent patterns of inequality are perpetuated through state policy-making. He does, however, argue that the state operates independently of key economic actors, and it is this assertion which has proved problematic in his theory of corporatism since it fails to take account of the system of economic constraints, national and international, within which governments must continue to operate. To put it simply, there is precious little evidence to suggest that recent British governments have controlled and directed key sectors of the British economy.

In his recent book, which reports a case study of the post-war planning process in Oxford, Simmie (1981) rejects this notion of corporatism as involving state direction of the economy, and argues instead that corporatism must be understood to refer to a system of inter-organisational bargaining conducted between interests which are incorporated within the state policy-making process. This definition, which is broadly consistent with the formulations offered by Cawson (1977), Jessop (1979) and others, differs from that developed by Pahl and Winkler in that its focus is on how different groups may influence state policy-making rather than on how the state may direct the activities of different economic interests in society. Such a definition is, I believe, potentially very fruitful. The problem with Simmie's work, however, is that he tries to make the concept do too much. In particular, he tries to use it against marxist theory rather than in conjunction with it. Simmie's basic argument is that, in advanced capitalist societies where the market has to a large extent been eroded by state intervention, class is no longer an appropriate concept for analysing political processes. Power, he suggests, is now a function not of class but of organisation.

The problem with this argument is that it reifies the concept of class by assuming that classes are empirical 'things' which used to play an important role in politics but which now have been replaced by other 'things' called organistaions. This would seem to miss the point of a marxist class analysis of politics; namely that classes are constituted through their organised practices and that class struggle becomes manifest in economic and political organisations such as trade unions, political parties, corporate bureaucracies and so on. To say that "on the one hand there are social classes" and "on the other hand there are organisations" (1981:54-5) is meaningless, for the problem (and as we saw earlier, it is a problem which marxist political theories have been unable to resolve) is precisely how and in what sense do political organisations 'represent' class interests? To deny absolutely the relevance of the concept of class is simply to side-step the issue, for we need to establish empirically and theoretically the relation, if any, between class interests and various organistaional forms. To say that policies are determined through bargaining between organisations is to say very little, for the question surely is which interests do the successful organisations represent?

There is, I would imagine, considerable evidence from Simmie's own case study to suggest that capitalist class interests, far from being irrelevant to an analysis of the political process, remain central to it. Thus he shows how commercial and industrial firms were able to exert pressure successfully in those issues where local planning decisions affected their interests, and he provides ample evidence to suggest that such firms were able to evade certain planning regulations while turning others to their advantage.

Such evidence does not, of course, warrant a marxist conclusion that the 'local state' is in some way representing the interests of local capital in the local political arena, but it does suggest (just as my own work in Croydon suggested) that in certain <u>types</u> of issues there may be a relation between economic (class) power and political power which it would be facile to deny. To repeat once more, such a conclusion is by no means incompatible with Weber's insistence on the analytical separation of economics and politics, since analytical autonomy does not rule out the possibility of a strong empirical relation between the two. The danger in Simmie's work is that, in his eagerness to develop an alternative to marxism, he ends up by denying the relevance of class power (economic domination) altogether.

Related to this is another danger, namely that of substituting one general political theory (that of marxism) with another (that of corporatism). Reading his case study of Oxford, one is struck by how little evidence there is to support his conclusion that there has been a "movement towards the incorporation of major private organisations into local planning authority decision-making" (291). Not only does Simmie himself admit that organised labour (surely one of the most important organised interests in a town at the centre of the British car industry) was virtually inactive in local politics, but the evidence of corporate influence which he does discuss suggests not the existence of an exclusive corporatist polity, but rather a system of imperfect pluralism in which different groups mobilise with greater or lesser effect at different times over different issues. As I have pointed out in relation to Pahl's work, pluralist political theory is distinct from corporatist theory and from weberian political sociology in general. It is all the more disconcerting, then, that at one point in his book, Simmie suggests that 'corporatism' and 'imperfect pluralism' are synonymous (106), for this is to render both concepts so general as to be virtually worthless. If a corporatist mode of interest mediation is relatively exclusive while a pluralist mode is relatively open, then it clearly makes no sense to define the one in terms of the other.

It is, however, only by employing the concept of corporatism in such a loose and broad fashion that Simmie is able to hold to his view that corporatist theory is, like the marxist theories which he seeks to replace, generalizable. In fact, his case study provides no evidence of the close, informal, regular and exclusive relations between the planners and private interests which are the hallmark of a corporatist polity. Nor is this surprising, for although there are signs that corporatist modes of policy-formation may be developing at the local level in Britain (the Urban Development Corporations are an obvious example), it is at national and regional levels of government that they are most fully developed.

This brings me on to the main point of this discussion which is that, rather than attempting to develop general theories of the state (be they relative autonomy theories or corporatist theories), it may be advisable to explore the possibility that political processes operate differently in different contexts and thus require different and complementary theories to explain them. It is, for example, possible to develop the hypothesis that, while local policy-making is typically the product of a process of imperfect pluralism in which different sectors of the local population compete over a range of different issues, national and regional policy-making is increasingly determined within a relatively exclusive sector of the polity in which the major class interests of the big capital and organised labour are directly represented. It could further be argued that different state functions tend to be located at these different levels of government such that the corporate sector at national and regional levels tends to determine policies affecting the management of the economy (for example, profit rates, interest rates, wage levels, grants and subsidies, taxation), while the pluralist competition at local level tends to revolve around issues of social consumption (for example, housing, education, welfare). Such a formulation could also help to resolve the problem concerning the relation between class and politics which is unresolved in marxist theory and sidestepped in Simmie's work. It would suggest that political struggles can be analysed in class terms to the extent that they arise over those aspects of state intervention which directly affect the relation between capital and labour in the workplace, while political conflicts which arise over the distribution of social consumption in the community tend rather to be constituted on the basis of what Dunleavy has termed 'sectoral alignments' as opposed to classes.

I have developed these arguments in much more detail elsewhere (see especially Saunders, 1980 and 1981a) and I do not seek to defend or enlarge upon them here except to note that such formulations can prove useful and fruitful in developing empirical work (for example, in the field of central-local government relations which is the area in which I hope to apply this framework in future research (see Saunders, 1982). But for the purposes of this paper, my own attempt to develop an analytical framework for the study of urban politics and to derive empirically-testable hypotheses from it is irrelevant save as an illustration of how Weber's methodology in general and political sociology in particular, may provide a rich source for sociological research on urban politics in the 1980s.

The methodology is important for its assertion of the centrality of ideal types for any sociological theory and research, and for its recognition of the inevitable partiality and one-sidedness of any sociological account. We would be well advised to give up the search for ultimate causes and total explanations, and to concentrate instead on identifying some of the factors which help to explain some aspects of contemporary problems and issues. To do this, we must delimit the scope of our analyses and recognise the limitations of our theories, and this will involve the construction of ideal typical analytical frameworks. It is no more realistic to expect a corporatist theory of politics to be universally and generally applicable than it is a class theory. It is therefore important to begin to specify in which situations which theories are likely to be appropriate.

Political sociology is important for its emphasis on the autonomy of economics and politics. This should not, however, be confused with any notion of the state operating independently of economic forces (as in Pahl) or neutrally between classes (as in pluralist theory). All that the autonomy of the economic and the political means is that the relationship between them remains to be discovered through empirical research and cannot be deduced through prior theorising. Theoretical speculation in the field of urban political studies has far outrun our empirical knowledge of how state policies come to be determined in different situations and in relation to different areas of intervention. It is my belief that in the sociology of Max Weber we can find some useful tools with which to begin developing such knowledge. The time for self-indulgent theoreticism is past; it is time we got our hands dirty.

NOTE

This is a revised and condensed version of a paper originally entitled 'Is there a weberian urban sociology', which was given at seminars at the University of Melbourne Department of Town and Regional Planning, and at the Urban Research Unit at the Australian National University in July and August 1981. The paper was later substantially revised and presented at the annual conference of the Institute of British Geographers at the University of Southampton in January 1982. This final version owes much to the comments and criticisms received on these occasions, and I should particularly like to thank Peter Williams and Chris Paris for their extended written comments on earlier drafts.

REFERENCES

Althusser, L. (1969) For Marx. London, Allen Lane.
Bhaskar, R. (1975) A Realist Theory of Science. Leeds, Books Publishing Co.
Birch, A. (1964) Representative and Responsible Government. London, Allen and Unwin.
Castells, M. (1977) The Urban Question. London, Edward Arnold.
Castells, M. (1978) City Class and Power. London, Macmillan.
Cawson, A. (1979) 'Environmental Planning and the Politics of Corporatism', Urban and Regional Studies Working Papers. 7, University of Sussex.
Dahl, R. (1956) A Preface to Democratic Theory. Chicago, Chicago University Press.
Dickens, P. (1977) 'Social change, housing and the state', in Harloe, M., ed., Proceedings of the conference on urban change and conflict. London, Centre for Environmental Studies.
Dunleavy, P. (1980) Urban Political Analysis. London, Macmillan.
Goldsmith, M. (1980) Politics, Planning and the City. London, Hutchinson.
Harloe, M. (1979) 'Marxism, the state and the urban question', in Crouch, G., ed., State and Economy in Contemporary Capitalism. London, Croom Helm.
Hill, R. (1977)'Two divergent theories of the state', Internatio Journal of Urban and Regional Research, 1, 37-44.
Hindess, B. (1977) Philosophy and Methodology in the Social Sciences. Brighton, Harvester.
Hirsch, J. (1978) 'The state apparatus and social reproduction', in Holloway, J. and Picciotto, S., eds., State and Capital. London, Edward Arnold.
Hirst, P. (1977) 'Economic classes and politics' in Hunt, A. ed., Class and Class Structure. London, Lawrence and Wishart.
Hirst, P. (1979) On Law and Ideology. London, Macmillan.
Hooper, A. (1982) 'Neo-Weberian political sociology: the politics of collective consumption', Political Geography Quarterly, 1, 1:97-107.
Jessop, B. (1977) 'Recent theories of the Capitalist State', Cambridge Journal of Economics, 1:353-373.
Lambert, J., Paris, C. and Blackaby, B. (1978) Housing Policy and the State. London, Macmillan.
Lebas, E. (1982) 'Urban and regional sociology: a decade of marxist and critical approaches', Current Sociology, 30.1:(whole issue).
Marx, K. (1976) Capital, Volume 1. Harmondsworth, Penguin.
Marxism Today (1979) Interview with Nicos Poulantzas conducted by Stuart Hall and Allan Hunt, 23, July:194-202.
Mingione, E. (1981) Social Conflict and the City. Oxford, Blackwell.
Pahl, R. (1977a) 'Managers, technical experts and the state', in

Harloe, M., ed., Captive Cities. Wiley.
Pahl, R. (1977b) 'Stratification, the relation between states and urban and regional development' International Journal of Urban and Regional Research, 1, 1:6-17.
Pahl, R. (1979) 'Socio-political factors in resource allocation', in Herbert, D. and Smith, D., eds., Social problems and the City. Oxford, Oxford University Press.
Pahl, R. and Winkler, J. (1974) 'The Economic Elite : theory and practice', in Stanworth and Giddens (1974).
Payne, G., Dingwall, R., Payne, J. and Carter, M. (1981) Sociology and Social Research. London, Routledge and Kegan Paul.
Pickvance, C. (1977) 'Marxist approaches to the study of urban politics', International Journal of Urban and Regional Research, 1,2218-55.
Saunders, P. (1980) 'Towards a non-spatial urban sociology' Urban and Regional Studies Working Papers, No.21, University of Sussex.
Saunders, P. (1981a) Social theory and the urban question. London, Hutchinson.
Saunders, P. (1981b) 'The crisis of local government in Melbourne' University of Melbourne Planning Papers, Melbourne.
Saunders, P. (1982) 'Why study central-local relations?', Local Government Studies, 8, 2:55-66.
Sayer, A. (1979) 'Theory and empirical research in urban and regional political economy', Urban and Regional Studies Working Papers, No.14, University of Sussex.
Sayer, A. (1980) 'Response' included in Saunders, 1980.
Sayer, D. (1979) Marx's Method : ideology, science and critique in 'Capital', Brighton, Harvester Press.
Simmie, J. (1981) Power, Property and Corporatism. London, Macmillan.
Stanworth, P. and Giddens, A., eds., (1974) Elites and power in British Society. Cambridge, Cambridge University Press.
Weber, M. (1949) The Methodology of the Social Sciences. Glencoe Free Press.
Weber, M. (1968) Economy and Society, New York, Bedminster Press.

4 Women and suburban housing

Post-war planning in Sydney, 1943-61

Carolyn Allport

The commitment to reconstruction planning, begun during the early 1940s in Australia, came from four distinct sources: the legacy of the depression, an experience that many groups in the community but particularly the labour movement were determined would not be repeated; social reform movements of the late 1930s, usually middle-class led, which had seen the growth in New South Wales and Victoria of slum abolition groups and the mammoth investigatory work of the Federal Government's Joint Committee on Social Security; the military need for morale on the home front; and the practicalities of planning for demobilisation after war which was intimately tied to the reassertion of women's traditional role as unpaid housekeepers and baby-minders. The reconstruction debate, spearheaded by the Commonwealth Government's Department of Post-War Reconstruction and paralleled by State Government departments and a plethora of community groups, did review almost all aspects of economic and social life within the context of the growing acceptance of central government planning. The canvas was indeed wide. Yet the planning of national policies was complicated by the international negotiations concerning full employment, tariff reductions and international monetary stabilisation, and the assertion of economic superiority by the United States. Not the least troublesome was the thorny question of Commonwealth-State functions and relations.

For students of the city, such reconstruction plans are particularly important. The establishment of the Commonwealth Housing Commission and publication of its Report in 1944 was crucial since it brought the focus of city reform, town planning and slum clearance within the parameters of financial commitment by both State and Federal governments. Further, the durability of its major policy recommendation - the funding, by the Commonwealth, of low-cost housing for low-income groups to be built and administered by State Housing Commissions with the provision that such funding should be coordinated with town planning policy - has transformed the structure and design of post-war housing in Australia. Although the Commission's Final Report recommended that all poor people should be assisted with housing, the State Housing Commissions' programmes have been shown to have 'helped only a small proportion of these people while it helps many people who have not remained poor' (Neutze, 1981:108).

Certainly it can be stated that the Commonwealth Report had hoped to reduce the inequities associated with the housing market, chiefly by favouring government intervention at the lower end of the rental market, and to relieve the massive housing shortage by effective government planning. The belief in planning as the universal panacea was strong among those associated with the Commonwealth Reconstruction Department, and to a certain extent among many involved in welfare work in Sydney. What was common to them all was a belief that the unplanned city growth of the preceding decades had resulted in profound social and economic diseconomies for the majority of residents. Like the town planning movement of the turn of the century, poorer housing of the city was now seen as an eyesore, an indictment of a modern society. Traffic congestion was both an inconvenience and a drain on the efficiency of the city economy, particularly because it lengthened the journey to work. Women, if they had a place in these discussions, were seen almost solely within their domestic role. Although women had worked in great numbers at a variety of occupations during the war, it was assumed that they would return to either the traditional light-manufacturing employment, or to the private home. In the minds of those discussing the plans for the post-war world, sex roles were clearly and sharply differentiated, even if the reality was not. Rather than articulate a post-war role for women based on the changing skills and employment patterns of the war years, those involved with the reconstruction planning harkened back to the pattern of domesticity established in the 1920s. With declining marital fertility, a rapid rise in marriage rates and a falling participation rate for the female work force, the role of women as mothers and homemakers was enhanced during the inter-war period. Pringle (1979:7) argues that the high social status given to homemaking in the 1920s was reflected in the setting up of Housewives and Country Women's Associations but that homemaking did not match the social status of most male occupations. Within this pattern, the depression of the 1930s and the profound social changes which accompanied the war were seen as discontinuities to the established homemaker domestic role. The return to normalcy for women in the eyes of planners thus ignored the direct experience of women in the preceding years, even though the Reconstruction Department's own surveys showed clearly that almost 70 per cent of servicewomen desired paid employment after the war, many in the new areas of work opened up by the war (Sydney Morning Herald March 14, 1944).

Central to both the Commonwealth Housing Commission's suggestions and to the later County of Cumberland Planning Scheme, was the necessity to decentralise both population and economic activity from the over-crowded central city. The shocking condition of much of the sub-standard housing investigated by both bodies, the continuing work of the slum-abolition groups, the influence of both English and American town-planning ideas and the availability of cheap suburban land, all led to a strong environmental determinist belief in the virtues of suburban living. Under the influence of the Greater London Plan of 1944 and its emphasis on the human geography of planning, the Cumberland Plan was emphatic that suburban sprawl must not go unchecked. Future suburban developments should be part of an integrated plan for the whole Sydney region. Working and living areas were to be separated yet conveniently located within each district linked by a green belt of open space.

Overlaying these planning developments, was the far more complex ideology of home ownership which, among middle-class reform groups and many parliamentarians, was justified on philosophical, military, moral and socio-economic terms. For those parliamentarians on the Left, the justification was more likely to be a reaction to the exploitation of many working class families by inner-city landlords. Again, such cries were not new, they had been heard throughout the twentieth century, but the experience of the Depression and war gave them a ferocity and immediacy not previously parallelled. The boom conditions following the war meant that there were many more women and men of Sydney, particulary among the working class, who could allow themselves to dream of a house of their own in some pretty little suburb. Moreover there were many more who could work consciously towards the fulfillment of that dream.

The essential focus of this work is the planning for and the process of suburbanisation in the decade or so after the war in Sydney, particularly as it effected women. Central to any attempt to assess the impact of urban planning for women, is a focus on the domestic relations of production, the gender pattern of spatial inequality of access to urban services and resources, and the relationship of urban plannng to such relations and inequalities.

THE COMMONWEALTH HOUSING COMMISSION - A NEW DEAL FOR THE OVER WORKED HOUSEWIFE

Potentially one of the most radicalising influences of Federal interest in urban planning during the war was the appointment of a woman to the Commonwealth Housing Commission, as housing was a task on which it was felt it was specially important to obtain women's views. Although it can be argued that this appointment did in fact restrict women's participation in the general reconstruction debate, and most importantly in the policy formulations for reconstruction, it did represent an acceptance that the experience and needs of women in the urban environment did differ from that of men. The appointment to the Commonwealth Housing Commission of a housewife, Mary Margaret Ryan, with no formal training in urban planning, in order to focus the Commission's work on the position of women in the domestic workplace, provides a fascinating example of both the attitudes of the patriarchal state towards women in the urban environment, and of the relationship between planning authorities, ideas and aspirations and the general community.

There had been great interest in the housing question among the women's organisations since the appointment of Flora Eldershaw to the earlier Division of Reconstruction Planning in 1941. Many of these organisations had given evidence before the Commonwealth Housing Commission, sponsored research and held conferences on housing and its relation to improving the status of women. Such interest and enthusiasm culminated in the Australian Women's Charter Conference held in November 1943, the recommendations of which echoed those set out in the Interim Report of the Commonwealth Housing Commission. These favoured town planning, slum clearance, decentralization and community centres. They called for a national plan to establish minimum standards, and an attack on the critical housing shortage. the appointment of women to housing schemes. Specifically, the Commonwealth Housing Commission perceived that the situation of the housewife would be improved if she was in some way released from the never-ending drudgery of housework and if her semi-isolation in the home could be relieved by integrating women more into the community.

The concern for the drudgery of housework came not so much from an outrage about the deplorable conditions in which most women worked in the private home, but from a concern with the falling birth rate. Mary Ryan herself, was convinced that Australia needed a high population growth if it was to survive in the post-war world; on the other hand, she found it quite explicable that women preferred smaller families to avoid the unending grind and worry that came from large families. In order to encourage women to have more children, housing standards must be improved and community support services provided to assist the mothers of young children. New streamlined modern homes which eliminated much of the drudgery of housework were felt by Ryan to be 'one way to make mothers more reconciled to remaining at their posts'(Ryan, 1944, Book IV:11).

Central to any attempt to improve the working conditions of housewives was the need for the application of modern scientific equipment to the home at a price within the reach of the low income-earner. The improvements foreseen by those giving evidence before the Commonwealth Housing Commission concentrated heavily on the electrification of domestic homes; the transformation of the kitchen into an efficient working area by the streamlining of furniture, fittings and the inclusion of essentials such as piped hot water; the integration of the kitchen and the living area; and the application of machinery to the laundry area.

From a study of the evidence and the work of the Commission, as well as the response of women to the Department of Post-War Reconstruction's community talks and the women's magazines, it is clear that there was a groundswell of opinion that these improvements must be made. In 1943, the Australian Womens Weekly asked its readers for suggestions for the new post-war home. It received thousands of letters which contained 'a withering barrage of criticisms against hideous, uncomfortable, inconvenient homes with dingy kitchens and bathrooms, dust-catching corners and poky rooms' (Australian Womens Weekly, May 8 1943:31).

Plans for the new post-war kitchen most captured the imagination of women. Billed as 'the modern American kitchen', it appeared in women's magazines, in literature of the gas and electricity companies, on countless newsreels and in publicity designed to excite and inspire women for their post-war role. The modern kitchen aimed to release the housewife from the uninteresting routine of domestic work by making the home more convenient to run and easing the preparation of meals. It envisaged a sink, stove, refrigerator and cupboards all built into one fitting so that the housewife could move from one appliance to another with a minimum of steps. Ovens and washtroughs were to be raised to a height sufficient to eliminate frequent back-breaking bending, and unnecessary walking would be saved by built-in dining alcoves for informal family meals. (Commonwealth Housing Commission, 1944:124). Such changes were also echoed in the advertisements for kitchen gadgetry, largely American, which populated the women's press during the 1950s and early 1960s.

Although the Commission received many submissions which suggested cost reductions through mass production of standardised equipment, they remained unconvinced that costs would fall sufficiently to allow the inclusion of refrigeration and hot water services in its minimum standard for government-sponsored dwellings, although it did suggest that both were essential. (Commonwealth Housing Commission, 1944:90)

All those involved in such reformulations of domestic working conditions saw technology as reducing the labour intensity and the time spent in housework. The invocation of the word 'modern' to stand for a whole new world of better things after the war assisted women to believe that technology could remove the drudgery of housework. Certainly it is clear that technology has transformed some aspects of housework from heavy to light production (for example hand washing and copper boiling as compared to modern washing machines) but it is less clear whether technology has reduced the time spent in housework. To the contrary, all available research, particularly from the United States where technology and electrification entered the housework arena earlier, suggests that the time spent on housework actually increased. Cowan (1974) compares the situations of rural and urban housewives in the United States in 1930 and 1947 and shows that in 1930, there was almost no difference in the time spent on housework (both groups reported around 60 hours per week). Yet in 1947 when city housewives had the benefit of electrification and many labour saving devices, their time spent on housework had risen to 78 hours per week in cities under 100,000 and 80 hours per week in larger cities, while rural housewives reported little change (Cowan, 1974:248). She further postulates that housewives have used labour-saving devices to raise the standard of household care (for example, daily washing with a washing machine as compared to weekly washing by hand and copper). Vanek (1974:117) supports Cowan's findings and traces the surveys beyond 1947 to show that the traditional housework tasks such as laundering and the acquistion of household goods and services takes up a much higher proportion of women's time than it used to. Pringle's (1979) work suggests that the qualitative changes in housework as a result of technology also meant increased time spent in child-care, reflecting the medical orthodoxy of maternal deprivation,

> 'Maternal deprivation' was perceived to have such disasterous effects that women were exhorted to attend to their children's needs every minute of the day, to be 'good mothers'. Mothers were encouraged to live vicariously through their children, ostensibly giving them the things which they themselves had not had. (Pringle, 1979:9)

Such expressions of structural change in the nature of housework fitted neatly with the concern among social reformers and planners of the 1940s with juvenile delinquency, and with the more conservative view, popular amongst church authorities, that women's war work was breaking down their exclusive role as full-time, unpaid homemakers and resulting in high levels of anti-social behaviour amongst children. Churchmen laid the blame for such anti-social behaviour at the feet of women.

While technology was to be the saviour of the housewife in the kitchen, improved physical support services were also necessary to assist women to become more integrated into the community and to reduce the semi-isolation of the housewives' home-centred existence. In 1943, Mona Ravenscroft, an anthropologist from Sydney University and an active member of the Union of Australian Women's (U.A.W.) housing research group, warned that the new housing schemes would need to take into consideration the different work environment that many women had experienced during the war. She pleaded that if suburban sprawl was to continue, then it must be balanced by adequate cultural outlets for the women who had returned to the home (Ravenscroft 1943:48).

Enthusiasm for community based services manifested itself in the community centre movement. The idea of community centres was immensely popular with those working in the Commonwealth Commission, the State Housing Commissions, particularly in Victoria, architects and town planning advocates, as well as many involved in social welfare organisations. The community centres were intended to provide for recreational, educational and social activities and to act as a focus for the provision of health services to the neighbourhood. With its base in the neighbourhood, the community centre fitted well into neighbourhood-district-region planning scheme of the Cumberland Plan, and was supported in the planning documents by lavishly illustrated drawings of centres which like many others of the time, combined swimming pools with libraries, whist games with gymnasia, and public auditoria with kindergartens (Cumberland County Council 1948:115-116).

Associated with most planned community centres was the establishment of child-care centres where mothers could leave their children in the secure knowledge that they would be adequately cared for in their absence. While there was undoubtedly agreement between the reconstruction planners and the community on the advisability of providing community facilities (particularly since in those heady planning days of 1943 to mid 1944 no-one was talking about finance) the question of the provision of child-care centres (kindergartens) met with less than unanimous support. The majority of the women's organisations, with the exception of the more conservative ones such as the Mothers' Union, favoured the establishment of kindergartens either separately or within a community centre complex. Many of the middle class members of these organisations had been involved in the wartime creche system and were aware of the many difficulties that had occurred with the inadequate provision of child care services for women war workers. The Kindergaten Union, the Women's Services Guild, the Union of Australian Women, and the Care of the Child in Wartime Committee all requested that the Federal Government take responsibility for the pre-school child and called for increased government spending on kindergarten facilities (Australian Archives CP 43/1 43/101; CP 43/1 44/333).

There was however substantial opposition to the kindergarten movement primarily from the established churches, on the grounds 'that it undermined the family, the basis and unity of society' (Spearritt, 1974:594). The Community Facilities Committee when discussing this question felt that opposition to kindergartens was too strong for it to be included in their listing of the minimum standard of community facilities to be located within reasonable distance from every house (Community Facilities Committee, 1944:4). Their reaction was particularly unfortunate considering the public response to the Department of Post-War Reconstruction's own Housing Exhibition reported to have attracted over 100,000 people in Sydney in 1944. A questionnaire distributed at the Exhibition revealed that 89 per cent of visitors believed full-day kindergartens were necessary (Sydney Morning Herald August 28 1944:4). Certainly, political pressure to avoid confrontation in the run up to the 1944 Powers Referendum was being exerted on all those working in post-war reconstruction. Still, in bowing down to the pressure against kindergartens, the Committee ignored one of the women's most important demands. Not only was the situation of paid working women placed in jeopardy but the hope of housewives to become more integrated with the community was also dealt a severe blow. Even further restrictions were placed on the standard of facilities recommended by the Committee with the withdrawal of any promises of federal financial aid (Treasury, 1945).

Unfortunately, with the withdrawal of federal finance, many local councils found community centre schemes beyond their financial capacity. Decentralised facilities such as baby health centres, local libraries or swimming pools found more favour and were economically feasible within the budget of local councils. Such decentralisation robbed women of the opportunities that the centres had offered in adult education and recreation pursuits. In areas of post-war suburban Sydney the only viable leisure pursuit for women became the local tennis-court, usually attached to a house, which charged low fees for a morning game and didn't mind small children running around. The withdrawal of support for kindergartens and the Treasury's opposition to Commonwealth financing dealt a severe blow to those low-income municipalities such as Redfern in Sydney which had expressed interest in the work of the Community Facilities Committee and had hoped that the government would build community centre complexes.

The participation of women on the Housing Commission did lead to a reconsideration of the working conditions of housewives. Certainly the demands for the rehabilitation of sub-standard housing, the introduction of labour-saving equipment, the redesigning of domestic work areas, and the provision of community support and extension services could have offered a revolutionary change for these women. In the final analysis, the position of women, whether working in the paid workforce or working in the home, was not considered important enough to outweigh other exigencies and the views of vested interest groups as to the appropriate role women should play in the new post-war society.

Labour-saving devices such as electrification of domestic appliances and the introduction of new appliances were considered to be too expensive to be recommended as a desirable minimum standard. Kindergartens and creches were considered to be too difficult a social question, subject to too much opposition, though not from women. Community facilities such as community centres which could have acted as a vehicle for adult education and adult recreation were considered too expensive by Treasury. It is important to recognise however that part of each of these movements for improved housing and working conditions for the housewife was inextricably linked to the propaganda for a reaffirmation of the family as the cornerstone of society and for women to see their return to normalcy as a return to the home. The dreams of the post-war home with its sparkling new American features and a kitchen equipped with all that the modern housewife could desire were considered to be the most effective way of encouraging women to breed more.

POST-WAR REALITY

While reconstruction planning had publicly raised the question of the situation of housewives and had been particularly critical of their working environment, it offered only dreams in answer to women's needs. The ensuing reality was very different. In order to examine whether these early planning ideas, and the operation of the later Cumberland Plan, made an impact on the working lives of women in post-war Sydney, one needs to adopt vastly different perspectives from those previously used in urban studies. Recent issues of the International Journal of Urban and Regional Research(1978) and Signs: Journal of Women in Culture and Society(1980) devoted to 'Women and the City' reinforce the methodological and research necessity of viewing women in the urban environment in a more specialised context. Even where women articulate the same urban needs as men, such as in the journey to work, the sexual division of labour within the home has often made the expression of their journey to work needs very different in form and structure to that of men.

At the general level, women's place in the urban system is clearly derived from the sexual division of labour which places the house/home as the essential locus of the domestic relations of production. Since the major proportion of women's domestic labour is structured through the system of marriage and the payment for such labour is indirect, Gamarnikow (1978:395-96) asserts that women's relation to the capital of the urban system is one of double dependence - on the husband as controller and appropriator of labour power and the male wage as the material conditions of capitalist domestic labour. Such dependence is then reflected and reinforced by the sexual dichotomies in the labour market and the low wages and status accorded to women's employment. Even though the Australian Women's Charter Conference, representing over ninety women's organizations, made the call in 1943 for a wage for the wife/housekeeper/mother and for equal pay for women, the patriarchal state in Australia has been loath to enter into the field of the domestic relations of production. Quite the contrary, the state in Australia has acted, with few recent exceptions, to reinforce such economic dependence and power relationships.

Central to the definition of such domestic relations of production, and of vital relevance to urban planning is the equation of 'house' with 'home' and the placing of home as the pivot of an idealised domestic community sharply separated from work. In their seminal essay, 'Landscape with Figures: Home and Community in English Society', Davidoff, L'Esperance and Newby (1976:139-175) correctly argue that the 'beau-ideal', the twin exultation of the rural village community and the home, gained special importance as a reaction to nineteenth-century industrialisation and reached its most defined level in the ideology of middle-class suburbanisation. Such suburban ideology is seen as "the quest for an organic community; small, self-sufficient and sharply differentiated from the outside world while the house and garden became both the setting and symbol of the domestic community" (Dresser, 1978:559).

Such separation of home and workplace has formed a central tenet of town planning ideology since its modern beginnings with Howard (1902) and others, and is mirrored in the Cumberland County Council Plan for post-war Sydney. Further, such geographic segregation intensifies the privatised nature of many women's lives and their exclusion from the public world. Saegert

(1980) argues that such long-standing beliefs have meant that the city and the suburb have symbolic gender associations. The city and urban life is equated with men - aggressive, assertive, definers of important world events, intellectual, powerful, active and sometimes dangerous. While suburban life is akin to women - domesticity, repose, closeness to nature, lack of seriousness, mindlessness and safety. That some present-day Sydney suburbs are very different from the 'domestic retreat' mentality so dominant in the suburbs of the 1920s and again in the 1950s and 1960s, owes as much to the changing economic and social roles of women and the expression of their changing urban demands as it does to any effective town planning. This is most particularly seen in the provision of localised child care schemes.

The powerful longevity of the domestic ideology not only sharply defined a division between the public nature of the city and the private nature of the suburb but also clearly restricted and delineated women's role within the privatised sector. It had immediate relevance to the middle-class woman of the nineteenth century - the 'angel of the house'. However, the pattern of working-class women's lives in the inner suburbs close to the city of Sydney and usually employed in the paid workforce meant that such women remained largely divorced from the domestic suburban ideology at least until the 1920s in Australia. With the opening up of suburban estates in the 1920s it was possible for more affluent sections of the working-class to join in the suburban home-ownership dream. However, home-ownership rates remained below 50 per cent and little real incursion was made into the pattern of inner-city rental housing characteristic of working-class housing in Australia (Spearritt, 1978:27-56). A decade and a half of depression and war pulled the 'house and garden' dream further away from such people. Yet the experience of evictions, homelessness and family dislocation during the Depression did harden the conviction for the dream of home ownership. Such feelings were intensified by the experience of wartime housing difficulties. Much of the housing stock had decayed through neglect, material shortages and wartime building restriction. New building in particular, was severely controlled, service personnel were given priority in rental accomodation and young marrieds usually lived with parents or relatives in the family home. Such policies produced conditions of gross overcrowding in many cases and privacy was rare. The dislocation produced by these housing restrictions further exacerbated the dislocations caused by the absence of family men on service duties.

By the time the Commonwealth Housing Commission presented its report, the housing shortage in Australia was acute - some 300,000 dwellings (Commonwealth Housing Commission, 1944:19). In Sydney alone, the shortage was estimated at 69,000 homes (Housing Commission of New South Wales, 1943:4). It is quite understandable then that in the post-war period with low interest rates, easy borrowing through War Service Loans, accumulated savings, lump sum cash payments for ex-service personnel, expansions in the lending policies of cooperative building societies and savings banks, the provision of low-cost rental, and after 1956, low-cost owner occupied housing by the State Housing Commissions, many more members of the working class could share in the suburban dream. Home ownership rates for Australia rose from 52.6 per cent (1947) to 63 per cent (1954) and to 70 per cent by 1961 (Kemeny, 1978:92).

Such developments had an important spatial consequence as large areas on the fringe of inter-war development jumped in population. Table 1 shows clearly the large decrease in residential population of the inner-city of Sydney and the corresponding growth of outer municipalities such as Ryde, Bankstown and Sutherland which although subdivided in the twenties, had remained mixed residential and market gardening areas throughout the 1930s and early 1940s.

While it is easy to understand people's desire for a home of their own, given the importance of the Depression and war experiences, it is more difficult to explain their preference for 'new' homes as opposed to the purchase of older homes in the city. There seem two important factors which taken with the rundown of the building stock since the 1930s, provide at least a partial answer. These are the psychological retreatism seen in many guises during the 1950s and the financial policies of the lending bodies.

It is often argued that the character of suburbanisation with its focus on domestic life produces insularity, particularly a retreat from more international, or even national issues in preference to a concentration on purely selfish parochial issues. It may well be the case that in the immediate post-war years, many young people were weary with the restrictions of war and with the dominance of the international situation caused by the War. Their return to normalcy may well have been articulated as an escape from the pressures, the horrors and the experiences of wartime - particularly for the men. As such a new start in an outer suburb with access to much country space, building a home in peace and serenity and being together for the first time in many years, often seemed the perfect way to put the war behind them. While such desires for domestic security were no doubt prevalent, one must also remember that divorce rates reached an all-time high rate of 11.6 per 10,000 of mean population in 1947 largely as a response to hasty wartime marriages and the pyschological problems of returning soldiers (Coughlan, 1957:116).

For most working class households moving out to these suburban areas it was the financial factor which most operated to induce their location choices and their preference for new buildings. Most lending institutions were prepared to lend between 75-90 per cent of the cost of a new home with the land as security, while a much larger deposit was required to purchase existing housing (Hill, 1959:141). Given that many working class households had some capital, their only real alternative was to purchase cheap

TABLE 1 Suburban Population in Sydney 1933-1954

	1933	Total %	1947	Total %	1954	Total %
Inner City	385,571	29.0	400,323	24.3	365,612	19.6
Eastern Suburbs	176,791	13.3	229,991	14.0	215,338	11.6
Illawarra	126,496	9.5	176,573	10.7	235,482	12.7
Canterbury-Bankstown	104,434	7.91	142,042	8.6	211,935	11.4
Inner Western	141,275	10.6	165,364	10.0	156,033	8.3
Outer Western	114,634	8.6	145,038	8.8	199,446	10.7
Fairfield-Liverpool	21,931	1.7	40,640	2.5	71,652	3.9
Northern Harbourside	97,544	7.3	119,255	7.2	117,087	6.3
Manly-Warringah	39,313	3.0	66,631	4.1	92,487	5.0
Kuring-gai-Willoughby	70,442	5.3	91,819	5.6	104,660	5.6
Ryde-Hornsby	50,971	3.3	68.196	4.2	91,953	4.9
Total	1,329,402	100.0	1,645,872	100.0	1,861,685	100.0

Source: Official Yearbook of New South Wales, 1957:63-64.

land in the outer suburbs and build their own home. Existing in an economic climate of shortages and efficient black markets, the owner-builder became a common feature of these suburban developments. Books, correspondence courses, various exhibitions and the proliferation of ready-cut building forms, all assisted the amateurs to build their own castles. By 1953 it was estimated that half of the new houses in New South Wales were being built by their owners (Official Yearbook of New South Wales 1957:680).

Suburban Women and Urban Access

To complete this assessment of the relationship of women to the suburban environment and urban planning a further series of issues must be considered. Focus must fall initially on women's access to housing. If such access is determined by marriage, then the question arises as to the degree of implicit discrimination which may apply to single, divorced or widowed women. Research conducted in the United States indicates that female-headed households, while being over-represented in welfare housing, are discriminated against by lending institutions (Werkerle, 1980:207). Certainly changes in headship rates amongst flat dwellers in Sydney indicate that women, whether single or married, are able to exercise access to this sector of the housing market, although the growth in headship rates is less for women than for men (Cardew, 1980:81). Nonetheless, in the immediate post-war period the institutional approach to housing assumed that women exercised their access through marriage. The rating system developed by the Commonwealth Housing Commission for the allocation of post-war dwellings identified female access through marriage and the family (including service widowhood). However, if an ex-service woman had combat zone experience or the equivalent then priority points were awarded, although less than for a man with equivalent experience. There is no mention of single women at all, even if those single women had dependants (Commonwealth Housing Commission, 1944:267). Similarly, suitability for a war-service loan although more liberal, necessitated that the applicant, whether serviceman or servicewoman, should satisfy the Director 'that he is married or is about to marry, or has dependants for whom it is necessary for him to maintain a home' (Hill, 1959:43). Both the Commonwealth Housing Commission and State Housing Commission systems favoured large families, yet of the 35,000 applicants to the New South Wales Housing Commission in 1946, only 30 per cent expressed demand for three-four bedroom houses, while 69 per cent demanded either small family units (39 per cent) or two-bedroom housing (30 per cent) (Housing Commission of New South Wales, 1946:25). Further, one must acknowledge that the Federal Government favoured the preferential granting of home loans to families with children; although it was never actually expressed as a distinct policy from the central bank, it was certainly a part of the political rhetoric throughout the 1950s and early 1960s.

Women's access to employment, transportation, leisure

activities and social services within the urban environment is critically connected to their domestic role and the support by the patriarchal state given to that role in urban and social planning. In a major study of the journey to work in Sydney, Manning has shown that "despite the apparent similarity in the overall availability of local employment for men and women, the travel patterns of the two sexes are quite different" (Manning 1978:78). In general, it was found that women travel shorter distances to work owing to a preference by married women for reduced time spent in travelling. The lower income received by women leads to an inability to pay for speedy travel, with more reliance on walking and bus modes (see Manning, 1978:80). Manning further claims that such transport preferences by women "implies that they choose nearer jobs than men at some sacrifice of the range of choices" (Manning, 1978:90). Studies of working class women in Melbourne support this tendency for married women in particular to prefer work close to home (Bryson and Thompson, 1972:56; Gepp, 1968). More importantly Bryson and Thompson's study shows that the reduced mobility of full-time working mothers and the availability of industrial work close to home resulted in a much greater proportion of factory work amongst this group as compared to single girls who were more concentrated in office work and who had wider geograpbhic mobility (Table 2).

TABLE 2 Occupations of Daughters and Mothers Working Full-time in 'Newtown' (1966)

	Daughters %	Mothers %
Industrial	33.3	50.0
Service	-	8.1
Sales	11.9	10.8
Office Work	50.0	27.0
Professional	4.8	4.1
	100.0	100.0

Source: Bryson and Thompson, 1972:57

More specific studies in the United States and France (the Paris region) support the theory that women choose to restrict the time spent on the journey to work in response to domestic responsibilities. In addition, research in the United States (Werkerle, 1980:106) suggests that such choice increases the "accessibility costs" of gaining employment and urban services, while the Paris study (Foder, 1978:463-481) suggests that women are subject to daily problems in co-ordinating commuting time and work hours with the operational hours of child-care centres.

It is obvious that access to transport and consequent questions of equity have an important relationship with the spatial allocation of the centres of production and consumption; this is particularly important when considering the changes wrought by corporate capitalism to the built environment in the post-war years. Changes in the production locus of the city have resulted in the decline of central-city manufacturing and the rise of peripherally located industrial estates, although some space-efficient enterprises such as printing, clothing, footwear and jewellery have remained in the city (Neutze, 1977:98). Associated with this trend has also been the rise in employment in office industries, particularly the female dominated clerical positions (Neutze, 1977:101).

Aided by the Cumberland County Plan whose zoning provisions precluded substantial expansion of industry in the inner areas of the city, central area factory jobs fell from 168,800 (67.7 per cent of metropolitan jobs) in 1945 to 149,800 (45.4per cent of metropolitan jobs) in 1961 (Alexander, 1981:152). Certainly the plan was successful in meeting its objective of a more equal distribution of jobs in relation to residences. Alexander shows that 'the ratio of manufacturing jobs to resident manufacturing workers increased in all of the designated suburban areas with the notable exception of the fringe district of Baulkham Hills-Blacktown where a good deal of greater than expected population increase occurred' (1981:153). Unfortunately such statistical information of employment-residence correlation has only been compiled in aggregate terms; there are no figures available for female workers. All indications are that the situation for women workers was very different from that of male workers. Manning's work clearly shows that in general female workers have shorter journeys to work and are more captive public transport users than men; yet the trend was clearly for female job opportunities to grow faster in the city centre. Further Black (1978:123) showed that the suburbanisation of factory jobs benefited the motorist to a greater degree than the user of public transport. The effects for women are clear - those women who wished to work close to home in the suburban districts were restricted in their job-search area and range of employment, since traditional female areas were increasing in the city-centre and restricted car access meant that many could not take advantage of manufacturing decentralisation. For many women who moved out to the western suburbs from the slum clearance schemes, such movement in the early years presented immense difficulties, particularly in respect to access to the traditional outwork of the clothing industry.

Such spatial changes in production and its effects on the female labour force were recognised in the Cumberland Plan (1948:95). While it did not specifically isolate the more general effects of the changing pattern of urbanisation on women, for example it did not include gender differences in its important statistical analysis of the employment/living/working relationship (1948:44), it did encourage decentralisation of factory employment partially on the basis of attracting married women to the labour scarce paid economy. It specifically stated that conditions of employment must be made attractive for women (1948:62). Although not entering the debate regarding how such attractiveness was to be achieved, the plan boldly advocated community support services such as community centres, kindergartens and local health centres. Such facilities together with a neighbourhood base for planning and a commitment to more circumferential transport services would have increased the access of suburban inhabitants, particularly women, to urban services previously concentrated in the city centre. For the woman who was solely restricted to the domestic situation, transport access may well have been one of deprivation. The diminishing importance of public transport in the post-war years and the rise of mass motor-car ownership (Spearritt 1978:168), affected the mobility of women and their access to urban services markedly. Firstly, although it can be claimed that the motor car can operate to increase access by enabling more direct cross-suburban travel in comparison to the city-centred public transport system, initially it operated to decrease transport access for women, particularly since most women during the 1950s did not drive. More generally, accessibility to motor car use has been shown to contain great disparities among men and women, different age groups and different physically handicapped groups (Morris, 1981:21).

Car ownership further encouraged a greater dispersal of residential settlement, often without any other mode of transport provided to service such settlements. The early sites for the New South Wales Housing Commission's building at Granville and Westmead were criticised for their lack of public transport. The Granville development was two miles from the nearest rail station and half-a-mile from the nearest bus service (Sydney Morning Herald, August 15 1944:2). Further, for a large part of the suburban population railway access was inconvenient for work and leisure journeys. New suburban developments were mainly serviced by private bus companies since the Department of Government Transport did not consider extension into new suburbs an economic proposition. The lack of feeder-bus services and the infrequency of private-bus services, at best they ran once an hour, further decreased alternative transport services (Jeans and Logan, 1961:39). According to many women living in the Bankstown area it was easier to walk, even if it did take much longer and restricted the area to a 2-3 mile radius (Allport, 1980).

Even when women did drive, they did not necessarily have access to the car to assist in the purchase of shopping. Jeans and Logan (1961:39) quote the example of the Seven Hills shopping centre which, at the start of the 1960s, had a large surplus of parking space and drew its trade from an area in which people walked or rode in buses. Nor did they have personal mobility, as the one-car household had to service the transport needs of all household members including children's leisure needs. It is clear from the numerous studies quoted by Manning (1981) that one of the basic assumptions of modern urban design - all adults have a car and children will be able to use motor transport for all trips apart from the journey to school - has resulted in considerable inequities in transport access, particularly in the one-car household. It is important to note that many leisure developments catering for children were built on low level land, unusable for residential development, and were not conveniently located for transport routes. Further, the numbers of children involved at any one time was often not sufficient for private bus services to cater to such areas. Consequently children needed to be transported both to after-school leisure activities and to weekend engagements. Thus women's access to car transport had to fit in with the primary need for the wage earner to get to work, the hours of various urban services such as baby health centres, doctors, dentists, shopping complexes and libraries, the arrival home of children from school and the need for leisure transport by these children.

Harman (forthcoming) argues that changes in consumption and social reproduction have meant that transport access for the suburban woman has become more important as consumption has become more socialised. Using the work of Mullins (1981) and Pringle (1981) she notes that shifts away from the self-sufficient domestic economy, "the demise of the Australian urban vegetable garden, fowl run, fruit trees, and 'do-it-yourself' skills in home repair, cooking and sewing" have resulted in an increase in the number and frequency of trips necessary to service the home. Reductions in home-delivery services, home visits by doctors, the decline of the corner-store and the growth of car-orientated supermarket complexes further increase the number of trips or vary multi-purpose trips. For the woman without a car, and perhaps accompanied by one or two pre-school children, such shopping expeditions can become a nightmare - her work not only involves an increased physical load to carry home but also child supervision in the shopping complex and on the journey home. Pringle believes that this takes up a much higher proportion of women's time than it used to and argues that the widening of women's consumption activities is a result of production being gradually removed from the home. Further the development of capitalist markets after the war "depended on the promotion of consumption as a 'way of life' and the superiority (and relative cheapness) of the 'bought' over the home-made" (Pringle, 1981:12). The organization of selling also acted to increase the labour and hours necessary to acquire household goods and services. By 1961-62 while self-service establishments constituted only 16 per cent of rental outlets, they held 71 per cent of the grocery trade (Johnston and Rimmer, 1969:41). Certainly as Dyer (1979:2) notes it is useful to list the regular visitors to a suburban home in 1938 and notice how few have survived - icemen, roaming vendors of fresh food, (butcher, greengrocer, grocer, rabbito), bottle-o, rag and bone-men, woodmen, clothes prop men, rent and insurance collectors.

While both Harman and Pringle provide analyses which are particularly perceptive, the actual timing of the process deserves more subtle analysis. Many home delivery services had been severely curtailed during the war, a fact Mary Ryan continually bemoaned; yet these were an important part of life in the pre-shop Housing Commission areas, particularly the temporary settlements. It is a contention of this work that the developments Harman and Pringle refer to did not permeate the lives of the majority of suburban women, particularly those in the fast growing outer-municipalities until the 1960s. The picture that emerges from the Bankstown study was that housework, including the production of home-consumed goods, was still heavily labour intensive until the 1960s and that the self-sufficient domestic economy was still strongly aided by the home-centredness found in many studies of suburban households.

Nonetheless, it is true that in the building of shopping centres on housing estates by the New South Wales Housing Commission, such centres were often inconveniently located for many with little transport access. The Commission did not permit the building of isolated corner shops; it favoured the small grouping of shops usually located in the centre of the estate, as at Lalor Park (Jeans and Logan, 1961:46). In some of the larger centres, such as Ermington or Villawood, a large shopping centre was built which attracted chain-stores which, with cheaper prices, offered some advantages to outweigh the transport difficulties. More often than not, only a small number of shops was provided "in small blocks unattractive to the chain-store group but sufficiently concentrated to make the shopping trip inconveniently long for many people on the estate" (Jeans and Logan, 1961:43).

It is a pity that the corner-shop development was halted during the late 1950s for as Dyer (1979:30) notes such corner-shops provided an important social outlet for housewives. It also meant a particularly disconcerting experience for women who had moved from the inner-city areas where they were dependent on the local network for friendship, support and short-term financing for household expenditure, and where shopping was part of that intimate and familiar network (See Tennant, 1939; also Park, 1949; Connell and Irving, 1980:298-300).

Access to transport also affects access to other urban services such as health care, educational resources and leisure facilities. In both the Commonwealth Housing Commission's Final Report and the Cumberland Plan, emphasis was placed on the necessity for localised and decentralised health care systems. The Cumberland Plan proposed that each neighbourhood would have clinic services and that in addition to expansion of existing hospitals, new district hospitals of 200 beds would be built in centres such as Fairfield, Bankstown, Sutherland and Hurstville. These district hospitals were to be supported by outpatient, ambulance and welfare services (Cumberland County Council, 1948:117,179-80). While each of the twenty district centres nominated in the plan were to have a district hospital, only five such centres contained a hospital by 1975 (Alexander, 1980:163).

Certainly there was a political commitment to the decentralisation of hospital and medical services; yet Donald

(1980) argues that centralisation is still the major problem facing the Health Commission and that the voluntary board of management system and the lack of a research and planning division in the Hospitals Commission, were the principal reasons for the failure of the decentralisation proposals. The voluntary board of management system required that before a hospital could be available, the time consuming process of arousing public interest, petitioning the government and overseeing the construction process had to take place. Donald correctly argues that such a process was difficult in new settlement areas and that uncertainty about the continuing availability of finance for major capital works during the 1950s excerbated the situation (Donald, 1980:78).

Perhaps most crucial to all suburban women was the overcentralisation of obsteric and paediatric services. The bulk of the demand for these services after the War lay in the newly developed suburbs. Overwhelmingly, settlement in those areas came from the newly married young people providing the baby boom noted by demographers of the period (Jeans and Logan, 1961:31). Yet most women from these suburbs travelled to the city centre to have their babies in the specialist hospitals clustered there, necessitating long journeys for family or friends visiting. The situation had not changed by 1974 as Table 3 illustrates.

TABLE 3 Supply of Gynaecologists, Obstetricians and Paediatricians by Region: Sydney Metropolitan Area 1974

Region	Gynaecologists and Obstetricians per 100,000 women aged 15-44	Paediatricians per 100,000 persons aged less than 6 years
Central	88	168
Northern	11	18
Western	13	5
South Western	20	2
Southern	10	13
Inner Western	37	23
Sydney Metropolitan Area	37	33

Source: Donald, 1980:69.

Such a pattern indicates that long transport journeys were required not only for the birth of the child, but for ante-natal care, examinations after birth, and for specialist attention for children. In addition, although the Baby Health Centre movement did receive impetus from the reconstruction planning debate, the opening of new centres did not to keep pace with population increases. By 1954, compared with the County average there were at least half as many babies again to each baby health centre in the Fairfield, Liverpool and Ryde districts and twice as many in Blacktown and Hornsby (Jeans and Logan, 1961:40).

This lack of health services in suburban areas was compounded by the low level of expenditure by the local governments on health in areas which in fact required rather high levels of expenditure due to the lack of proper sewers (Jeans and Logan, 1961:40). The dependence on tin-pans and septics throughout the 1950s and part of the 1960s often created additional work for the household, particularly in low lying areas in the wet weather. With the lack of kerbs and guttering in most of the suburbs and the poor state of the roads, many of the residents spent large sections of their time "clearing street and road gutters, tending incinerators, carting rubbish, disinfecting, maintaining or repairing inefficient lavatories" (Brennan, 1965:304). With husbands at work, many of these tasks had to be undertaken by the women.

These patterns of access to urban services and resources such as employment, transport, sewerage and health need also be parallelled with studies on access to educational resources and leisure facilities in order to fully review existing concepts of spatial inequality in terms of gender and the circumstances surrounding the domestic relations of production. Black's study (1978) of comparative accessibility in different Sydney suburbs showed that within the acceptable range of a bus journey, many areas still lacked jobs, and medical and recreational facilities. On the other hand, primary schools and high schools were reasonably distributed within walking or bus journey (Manning, 1980:52). It seems that the provision of educational facilities was considered a high priority by residents of new areas. Many women who did not work in the paid work force expended a great deal of energy into community pressure campaigns for a new primary school or district high school and the reaction by the authorities often meant quick success (Kelly, 1957:402-3). They were however slower to react with the provision of leisure facilities. In 1944, the State Government began subsidising local libraries. By June 1947, sixty local councils had adopted the Act, forty-seven libraries were in operation, of which eight were in Sydney and the suburbs. By June 1952, there were still only twenty-four municipal libraries in the metropolitan area while the number of councils stood at forty. (New South Wales Statistical Register, 1948 and 1953). The Cumberland Plan suggested that each district centre should have a theatre assembly hall, a large central library supporting smaller neighbourhood libraries and a district exhibition hall for art and museum exhibitions and lectures (Cumberland County Council, 1948:177).

The community centre scheme which had offered so much and had been greeted so enthusiastically during the war, did not prove feasible in the outer suburbs. The only evidence of a community centre being built in Sydney during this period was the one at Castlecrag, an area setttled by highly motivated middle class professionals who were able to assist in the financing of the centre. Undoubtedly it was the high cost of such centres which deterred local councils from building the centres themselves. The financial base of local councils in the outer suburban areas was restricted by low land values and their failure to borrow to finance infrastructure building. Similarly the State government was hamstrung by its inability to continually finance the capital works necessary for the social infrastructure of metropolitan growth. Community service schemes, while theoretically necessary, were placed low on the agenda of both governments. Further, Menzies had made it quite clear in 1951 that the Cumberland Planning Scheme could not expect financial support from the Commonwealth.

Intimately related to the question of access to urban services (as central components of the domestic relations of production) are the effects of housing, tenure types, architectural practices and zoning policies on housework and child care. Zoning in particular often limits the location of child care centres and commercial establishments which could act to make more communal aspects of housework such as laundering. Further, council ordinances and local government regulations often hinder the development of residential areas which are not based on the nuclear family. Building ordinances, if enforced, restrict communal living and only recently has legislation in New South Wales been amended to allow for extension of domestic homes by 'granny flats'.

Housework and child-care during the 1950s was essentially the sole preserve of women. The availability of overtime in the labour scarce economy, meant that men were absent from the home for long hours. Even traditional men's duties in child-care, such as the discipline of male children, fell to the women. Further many of these households were building their own homes and the extra earnings achieved through overtime were channelled into the building programme. Little money was available for the purchase of labour-saving household equipment envisaged by the Commonwealth Housing Commission. Many of the residents of the newly developed Bankstown area stated that the purchase of washing machines, electric refrigerators and floor polishers was delayed until the beginning of the 1960s. Women still used copper boilers for washing, although electrically powered, and spent one day a week on their knees polishing the bright, wooden floors. On wet days the women would need to dig makeshift drains to divert water in low-lying areas, and in their spare time (although few could remember having any) they harassed authorities for the establishment of schools, proper roads, effective street lighting and the improvement of local transport facilities (Allport, 1980).

Perhaps most alienating of all was the attitude of the local community towards those rehoused in Commission areas. Since the men were away so long at paid workplaces, it was the women who had to cope with the refusal of credit by local businesses, the cries of 'slum kids' at the local school, the cool reception at local church services and the paternalistic visits of Commission welfare officers. Such communities were a far cry from those envisaged by the Commonwealth Housing Commission.

REFERENCES

Alexander, I. (1981) 'Post War Metropolitan Planning: Goals and Realities,' in Troy, P., ed. Equity In The City. Sydney, Allen and Unwin.

Allport, C. (1980) Interviews by Carolyn Allport with residents at Revesby, Chester Hill and Bankstown (January-July, 1980).

Black, J. (1978) 'Changes to employment accessibility in Sydney', Papers, 3rd Meeting of the Australia and New Zealand Section, Regional Science Association, Melbourne.

Brennan, T. (1965) 'Urban Communities', in Davies, A. and Encel, S., eds., Australian Society. Melbourne, F.W. Cheshire.

Bryson, L. and Thompson, F. (1972) An Australian Newton: Life and Leadership in a new housing suburb. Ringwood, Penguin.

Cardew, R. (1980) 'Flats in Sydney: The Thirty Per Cent Solution', in Roe, J. ed. Twentieth Century Sydney Sydney, Hale and Iremonger.

Commonwealth Housing Commission (1944) Final Report.Sydney, Government Printer.

Community Facilities Committee, (1944) First Report. Canberra, Department of Post-War Reconstruction.

Cowan, R.S. (1974) 'A Case Study of Technological and Social Change: The Washing Machine and the Working Wife', in Hartman, M. and Banner, L.W., eds, Clio's Consciousness Raised. New York, Harper and Row.

Connell, R. and Irving, T. (1980) Class Structure in Australian History. Melbourne, Longman Cheshire.

Coughlan, W.G. (1957) 'Marriage Breakdown', in Elkan, A.P. ed., Marriage and the Family in Australia. Sydney, Angus and Robertson.

Cumberland County Council (1948) Report on the Planning Scheme for the County of Cumberland New South Wales. Sydney, Government Printer.

Davidoff, L., L'Esperance, J., Newby, H. (1976) 'Landscape with Figures: Home and Community in English Society', in Mitchell, J. and Oakley, A., eds., The Rights and Wrongs of Women. London, Penguin.

Donald, O.D. (1981) 'Medical Services' in Troy, P., ed. Equity in the City.

Dresser, M. (1978) "Review of 'Landscape with Figures'" International Journal of Urban and Regional Research, 2:558-563.

Dyer, S.W. (1979) 'Ordinary Suburban Thoughts', Paper presented at the Wrong Way - Go Back Conference, Sydney University 8-9 February.

Fodor, R. (1978) 'Day Care Policy in France and its Consequences for Women: A Study of the Metropolitan Paris Region', International Journal of Urban and Regional Research, 2:463-481.

Gamarnikow, E., (1978) Introduction to 'Women in the City', International Journal of Urban and Regional Research, 2:390-403.

Gepp, H. (1968) 'A Note on Working Wives', in Married Women in Industry: Three Surveys. Melbourne, Department of Labour and National Service.

Harman, E.J. (forthcoming) 'Capitalism, Patriarchy And The City', in Baldock, C. and Cass, B., eds, Women, Social Policy and the State in Australia. Sydney, Allen and Unwin.

Hill, M. (1959) Housing Finance in Australia. Melbourne, Melbourne University Press.

Housing Commission of New South Wales, Annual Reports.

Howard, E. (1902) Garden Cities of Tomorrow. Reprinted (1960), London, Faber and Faber.

Kelly, G. (1957) 'Portrait of a New Community', Meanjin, 16,4:399-407.

Kemeny, J. (1978) 'Home Ownership and Finance Capital', Journal of Australian Political Economy, 3:89-97.

Jeans, D., and Logan, M. (1961) 'The Problems of Growth in Sydney's New Suburbs' Australian Journal of Social Issues, 1, 1:29-48.

Johnston, R.J. and Rimmer, P.J. (1969) Retailing in Melbourne. Research School of Pacific Studies, Department of Human Geography Publication H9/3, Canberra, Australian National University.

Manning, I., (1978) The Journey to Work. Sydney, Allen and Unwin.

Manning, I. (1981) 'Traffic and Local Planning', in Troy, P., ed. Equity in the City.

Morris, J. (1981) 'Public Transport', in Troy, P., ed. Equity in the City

Mullins, P. (1981) 'Theoretical Perspectives on Australian Urbanization: 1. Material Components in the Reproduction of Australian Labour Power', Australian and New Zealand Journal of Sociology, 17,1:65-76.

Neutze, M. (1977) Urban Development in Australia. Sydney, Allen and Unwin.

Neutze, M. (1981) 'Housing' in Troy, P., ed. Equity in the City.

Park, R. (1949) Poor Man's Orange. Sydney, Angus and Robertson.

Pringle, R. (1979) 'Sexuality and the Suburban Dream', Australian and New Zealand Journal of Sociology, 15,2:4-15.

Pringle, R. (1981) 'Women and Consumer Capitalism', Mimeo School of Behavioural Sciences, Macquarie University.

Ravenscroft, M. (1943) 'The Housing Problem',Social Horizons, July.

Ryan, M., Unpublished Diaries, 1943-1944. (Held by the Family).

Saegert, S. (1980) 'Masculine Cities and Female Suburbs: Polarized Ideas, Contradictory Realities', Signs, 5, 3, Supplement:96-111.

Spearrit, P., (1974) 'The Kindergarten Movement', in Edgar, D.E., ed. Social Change in Australia: Readings in Sociology. Melbourne, Cheshire.

Spearritt, P. (1978) Sydney Since The Twenties. Sydney, Hale and Iremonger.
Tennant, K., (1959) Foveaux. Sydney, Angus and Robertson.
Treasury, (1945) Letter to the Director General, Post-War Reconstruction, 29/6/45, Australian Archives, CP 43/1 and 44/334.
Troy, P. ed. (1981) Equity in the City. Sydney, Allen and Unwin.
Vanek, J. (1974) 'Time Spent in Housework', Scientific American, November:116-20.
Wekerle, G. (1980) 'Women in the Urban Environment', Signs, 5, 3, Supplement:188-214.

5 Population mobility in Australia's urban areas

Chris Maher

The movement of population from one location to another is a basic dynamic of an urban system. Population mobility alters the relative distribution of individuals and households both between regions of a country and within areas of cities and towns and thus changes the nature of demand for housing, public and private finances and other services. As well, it acts to fulfil individual expectations and aspirations and to abet or frustrate goals in population distribution and resource provision. The study of population movements thus has a utility far beyond a mere description of where people live at a particular point in time, or the patterns of movement in space.

Evidence regarding the extent of population movement indicates that a significant proportion of the population is involved. The 1976 Census of Population and Dwellings for Australia showed that approximately 45 per cent of the population over five years of age resided in a different dwelling in 1976 to that of 1971. Turnover for the one year period 1975-76 was 17.3 per cent of the population. These figures are comparable to those of the United States, commonly regarded as the world's most mobile nation, where it is generally agreed that almost one person in five moves each year, some of course making multiple moves (Simmons, 1968), and almost half of the population (48 per cent) moved in the period 1970-75 (Quigley and Weinberg, 1977). These rates of movement are far in excess of a country such as Britain, where annual movement is estimated to be approximately 10 per cent (Short, 1978), and is even considerably higher than New Zealand where 14 per cent of the population moved between 1970 and 1971 (New Zealand Department of Statistics, 1976).

The reasons for movement vary widely. Some relate to choices exercised by a household; others can be related to factors external to the individual unit which imposes the need to move. Discretionary moves may be attributable to a search for improved economic circumstances, such as a new job or investment in housing; they may be related to environmental conditions or social aspirations; they can be a function of changing accessibility requirements of some individuals within a household; or they can be a reflection of changing accommodation needs of the entire household. Population mobility is not solely a function of household choice however. The actions of governments, institutions, firms, or other individuals may force the necessity of relocation upon a household, ranging from the displacement of population from dwellings expropriated to make room for public works, through to the allocation of households to public housing units, the arbitrary decision of a landlord to sell or to raise rents substantially, or to an individual being transferred within an employment situation.

Thus population mobility is a very complex issue. It is at once an effect of and a variable contributing to, economic, social, and spatial processes and outcomes. Regardless of the underlying reasons for movement, each contributes to an outcome where a population is redistributed over space. Not only is the outcome one where the popualtion may or may not be better served by the adjustments made to their location, but the process also affects the locations themselves. Changing levels of demand for housing, schooling, and other facilities are all products of a mobility process which have important implications for public policy involved in determining the extent of need for social facilities in various locations.

Unfortunately, knowledge of population movements in Australia is incomplete. In particular, little is known of the nature of mobility as it affects urban areas, largely because of the unavailability of sufficiently comprehensive data. Recent extensions to the Australian census however, particularly in 1976, have provided a means to extend our knowledge of the nature and extent of the process at a level less than that of relatively gross, aggregate flows. The purpose of this paper is to analyze the nature of population movements as they affect Australian urban areas, and to place the patterns, problems, and implications of these trends within the broader concerns of recent mobility literature.

MOBILITY AND MIGRATION: THE SCALE QUESTION

That population turnover is an important and active element of social change is well documented (Clark and Moore, 1980; Maher, 1982; Rossi, 1980; Short, 1976; Simmons, 1968). The processes and effects of population turnover are diverse and complex. Part of the complexity, as discussed above, results from the myriad of reasons behind population movement; part is a function of the variety of scales at which migration can occur. Immigrants newly arrived from overseas, households transferring from interstate, and individuals shifting from one flat to another within the same building are all identified as mobile. At any scale, part of the relocation process is attributable to the needs of the households to restructure their social and/or economic environment, and part stems from situations beyond the control of the population affected. Often, the immediate causes and effects of the various scales of movement are quite different, both quantitatively and qualitatively. In particular, a distinction can be made between movements occurring at an international or inter-regional scale, where the motivation is often economic; and those occurring at a much more localized scale within the confines of the same community and where the motivation is most frequently related to the adjustment of housing consumption. The former is generally referred to as migration; the latter as intra-urban mobility.

A distinction between the two scales has been made by Roseman (1971) when he argued that there were two types of adjustment involved, depending on the nature of the move. The first was what he termed a 'total displacement' move which describes the situation where the movers are subject to an alteration of their entire set of activities and contracts, including housing, jobs, schooling and social patterns. The concept applies particularly to those who move considerable distances, and describes the situation of inter-urban rural-urban and inter-country migrants. The second scale of move has been termed 'partial displacement' and describes a situation where only some of the activities and contracts are changed. Of these only housing is changed in all situations. What else is altered depends on the distance of the move. As most moves are over relatively short distances, schools, jobs, shopping habits and social contacts can remain the same. It is intra-urban movers who are generally those involved in partial displacements. Because of the fact that it is less disruptive, intra-urban mobility tends to be quantitatively more important, particularly in large urban areas.

Each component contributes to the major outcome of mobility - the redistribution of population from one area to another. At the broadest scale, immigration redistributes population among countries, with economic implications such as an increased labour force and a greater demand for goods and services, and with social implications relating to difficulties of adjusting to a foreign culture in the recipient country. Inter-regional movements, the other scale at which total displacement occurs, brings to the fore issues of economic opportunity. Migration at this scale is most responsive to variations in labour market performance, which in turn impinge upon the differential economic performance of regions in terms of growth, standard of living and a host of other economic indicators. At the lowest level, the partial displacements or population movements within urban areas also raise a number of social and economic concerns. Indices of housing market performance such as new buildings, vacancy rates, prices and rents are all affected by intra-urban mobility as are local needs for educational, health and welfare facilities.

THE CHANGING NATURE OF MOBILITY STUDIES

Despite a general consensus that the movement of households does represent a vital component of urban dynamics, the diversity and complexity of the process poses many conceptual and empirical problems. Because it is a readily identifiable and pervasive element of the urban scene, many have seen it as basic to an understanding of both individual and aggregate change. Yet while the process impinges on many issues such as housing and labour markets, individual welfare improvements, aggregate neighbourhood change, and so on, the problem is that it is central to few of these areas (Gale, 1980). There is currently a great deal of literature critical of many previous mobility studies. The difficulty, which has been encountered with much of the work to date, has been that it has been divorced from its interconnection with the broader set of historical or contextual factors. In particular, most work "has assumed rather than demonstrated that mobility is a significant subject of inquiry" (Harris and Moore, 1980:22). Thus while it has been argued that mobility is significant in relation to the operation of the housing market (Kendig, 1981; Quigley and Weinberg, 1977) the labour market (Cordey-Hayes and Gleave, 1974) and basic to an understanding of regional or neighbourhood population dynamics (Clark, 1976), it is becoming apparent that population movement is only one of a number of processes acting to influence both the distribution of population, and the range of opportunities, or 'life chances', faced by population groups in various locations. The nature of mobility itself, it is increasingly being argued, is a variable dependent upon broader factors relating to both political and institutional actions and policies, and to the general social and economic context within which the process is occurring.

Approaches to the relocation of population have evolved over time as the research focus has changed from a descriptive orientation through to a much greater concern with the relationships of migration and mobility to housing and labour markets, and to the role of movers as a dynamic element in urban and regional structure (Clark, 1981). Four separate themes can be identified. These are distinguishable by the extent of theoretical input and by the scale at which the analysis is conducted.

The first is what can be described as the analysis of event based information (Gale, 1978). In this approach, observations of actual movement are derived from either census or survey records, and are used to describe, summarize and model patterns of population turnover in a relatively simple fashion. Examples of this type of work are found in Moore (1969) and Simmons (1974). Issues such as the characteristics of the mobile population, the patterns of origins and destinations, and the impact of such movement on regions or neighbourhoods are all very relevant questions. A more recent concern with outcomes suggests that this approach will have a good deal to contribute to understanding the implications of aggregate moves on policy considerations such as housing and employment availability, service and facility distribution, and the differential access to resources that population dynamics imply.

The second broad approach to mobility studies (particularly those of intra-urban mobility) represented a turn away from description at the macro-scale to a search for explanantion through behavioural analysis. The most significant step in the behavioural approach was to change the scale of analysis from a consideration of broad trends, to an examination of the decision-making behaviour of the individual or household as the key to understanding urban change. The behaviouralists based their arguments on the assumption that observable patterns of population redistribution particularly in the intra-urban case were the result of individual decisions made in response to individual aspirations, perceptions and evaluation of the existing spatial structure. Best known in geographic literature is the two-stage model proposed by Brown and Moore (1970), where an initial decision to relocate, based on the achievement of a threshold level of dissatisfaction with the existing dwelling or location, is followed by the relocation decision where alternative dwellings and locations are assessed with respect to the aspiration region of the household or individual. Behavioural studies have used the concept of place utility to account for the manner in which choices are made between locations. While economists in particular have been critical of the fuzzy nature of this concept (Quigley and Weinberg, 1977), it has been retained in recent studies of mobility and the operation of the housing market (Hanushek and Quigley, 1978).

More recently, the behavioural approach has been criticized by those who feel that such individualistic explanations are insufficient to account for the magnitude and nature of population movement. Thus two other related but distinct sets of literature can be identified in recent discussions of mobility. One, which can be termed a structuralist approach, arises from the Western European marxist tradition of urban analysis which grew up at the end of the 1960s. General outlines of the approach are contained in Castells (1977), Harloe (1977), and Harvey (1975). It has found its immediate application in population turnover research through the work of British geographers such as Short (1976, 1978), Bassett and Short (1980), and Gray and Boddy (1979). This thrust has tended to concentrate on the theoretical analysis of structural forces acting both to determine variations in the range of choices available to a society stratified along class lines, and of factors important in moulding and altering aggregate urban patterns. Within these basic parameters, population movements are seen to be responses to the broader controls exerted by structural organization, as opposed to the behavioural approach which has seen mobility as the independent variable in spatial evolution.

The second approach has its origins primarily in North American literature on population movement. Here, concern has been expressed that attempts to establish a 'single theory of mobility' have not only failed but are based on incorrect assumptions about the nature of the process. The argument is made that,

> "...The dominant paradigm has emphasized micro-level analysis of household decision-making within a competitive market framework. It has focused on the development of general propositions about the relationship between

individual characteristics, preferences, and market mechanisms without regard to the constraints of specific socio-political environments". (Harris and Moore, 1980).

In the work of Moore in particular, there is a continuing theme that too often mobility studies are considered worthwhile solely because population turnover is a readily observable and common occurrence rather than because it can be related to wider social processes. However, this latter approach is considerably more optimistic about the worth of mobility studies, provided their broader antecedents are recognized, than is the structuralist view.

The two sets of literature, while having a common concern with placing the process within a broader framework of both historical and contextual factors, are differentiated largely in terms of scale. In the case of the British literature, this has consisted largely of theoretical work in a structuralist and marxist mode of analysis which has related to mobility studies through its analysis of the housing system. Thus, as Short (1978) argues, households are not autonomous decision-making units, and the behavioural aspects of population movement can be better explained as a form of adaptive behaviour to the system of housing supply and allocation which itself is dependent on the wider societal structure.

The American literature, on the other hand, has given much more attention to the influence of local contextual factors, particularly with reference to their implications for policy and planning initiatives with respect to housing. Indeed, Clark and Moore (1980) have been critical of the structuralist approach, arguing that while it discusses the ways in which household decisions are more the product of constraints rather than choice, the analysis is one-sided. Although it examines how housing needs and moves are created and mediated, they argue it says nothing of the subsequent effects of household movements either in individual or aggregate terms. Thus, in the approach being taken by Moore and Clark and others in the United States (Clark and Moore, 1978, 1980, forthcoming), there is much greater attention being paid to outcomes and their implications rather than to the nature of the relocation process itself.

The discussion of variations in approaches, contexts and scales at which population movement can be studied serves to demonstrate the difficulty which exists in attempting to conceptualize the process and to provide any single unified theory about mobility. It is a highly complex process related to individual aspirations and motivations, to variations in local contexts, and to the broad structural constraints of social, political and economic organization. Thus attempts to explain the nature and rates of mobility have tended to be partial, and at times contradictory. Because mobility is a component of so many facets of society such as housing markets, labour markets, investment opportunities, welfare considerations and a host of other criteria, it is highly unlikely that any single explanation or theory of mobility is likely to be derived.

At the same time, there are a number of issues arising from the mobility debate which warrant attention. In particular, a focus on the aggregate process and outcomes of population turnover can add to our knowledge of both how regional and local change occurs and the implications of these changes in terms of the dynamics of the urban system through the growth, decline or altered characteristics of urban centres, or local neighbourhoods. The remainder of this paper then concentrates on documenting the nature and rates of mobility in the context of Australian urban areas. The approach is primarily descriptive as so little is known about movement rates which affect urban areas. This analysis will however lead in the final section of the paper to an examination of the implications of the findings and will provide a basis on which alternative explanations may be explored.

MOVEMENT DATA FOR AUSTRALIAN RESEARCH

Population mobility studies have had some prominence in Australian urban and regional research for some time, although their scope has been greatly handicapped by inadequate data sources. Until the 1971 Census data on migration became available, most published work dealt with relatively limited surveys of specific urban areas. One of the earliest, and also one of the most ambitious, was the analysis of intra-urban mobility in Brisbane carried out by Moore (1969) based on changes of address traced through electoral rolls. Other Australian surveys have emphasized the behavioural aspects of the movement process, using utility or electoral roll data and follow-up interview surveys. Humphries (1973), Newton (1977), and Whitelaw and Robinson (1972) were all concerned to examine the nature of the reasons for, and the subsequent spatial patterns displayed by individual households in Newcastle, New South Wales, and Melbourne. A different approach, and one which attempted to examine the influence of supply factors, although again treating the housing system itself as a constant, was the work done on vacancy chains in the housing market in Melbourne (Maher, 1979), and the housing study by Kendig in Adelaide (Kendig, 1981). Little research has been carried out on the broader structural lines although the topic has been discussed (Bell, 1977; Kemeny, 1977; Kilmartin and Thorns, 1978).

Large scale, comprehensive studies of migration were not possible until the 1971 Census in Australia (Rowland and Mikkelson, 1979). In that census, a question was asked as to the residence of the respondent five years previously, in relation to present location. Data on movement for the 1966-71 period was then tabulated at the level of statistical division, which permitted the first large-scale overviews of the mobility of the Australian population (Cities Commission, 1974; de Castro Lopo, 1974; McKay and Whitelaw, 1977; Rowland, 1979). But because of the fact that moves were tabulated at a relatively large-scale (that of statistical division), no intra-urban or inter-urban moves could be analyzed other than at the level of the metropolitan statistical divisions. Thus more local impacts remained unexplored.

The scope of mobility data was considerably expanded in the 1976 census. The questions were phrased differently, and respondents were asked to nominate the actual address at which they had resided both one and five years previously. The tabulation of responses was then carried out at the level of local government area which enabled a much more detailed analysis of the spatial pattern, including the possibility of examining intra-urban mobility (Maher, forthcoming, a).

The addition of a further time scale (one year, 1975-76) as well as the five year period obtained initially in 1971 was a further extension of the data. Deriving information on mobility within a fixed period has the advantage of being a relatively simple question to answer. However, as the question is phrased only in terms of whether the individual is in the same dwelling as at the previous date, it is possible that multiple moves can be made within that specified period and hence the extent of mobility underestimated. The longer the period, the greater the probability of multiple moves being made and thus underestimation. The adoption of a one-year period does not eliminate the possibility of missing mutiple moves, but at least it has the effect of minimizing the problem.

The aggregate nature of the census data has both advantages and disadvantages. The information collated in the migration data files by the Australian Bureau of Statistics relates to the origins and destinations of movers, cross-tabulated by a number of measures of the social, demographic, and economic charcteristics of both movers and non-movers. Thus the data are suitable for macro-scale analysis of the broad trends in population redistribution, and provide the capacity to analyse in detail the outcomes in terms of the impact of mobility on regional and neighbourhood characteristics (Maher, forthcoming, b). They cannot contribute directly to an explanation of redistribution processes through either a micro-scale study of motivational aspects of the movers, or a broader structural analysis, although the implications of these can be raised. More significantly the data allows an examination of the impact of some of the local contextual variations on the mobility process (Maher, forthcoming, a).

One other proviso regarding the mobility question is that the measure of population movement really represents two types of movement. On the one hand, there is population turnover which is the extent to which a population is mobile under the assumption of a relatively static total population. On the other hand, the figures also include population growth, where individuals or households are moving in to an area and thus are recorded as not being in the same dwelling as previously, yet they are not replacing a pre-existing population but adding to it. This becomes important when examining the reasons for variations in mobility.

In order to analyse population movements both within and between urban areas, settlements above a minimum population size of 10,000 in 1976 were included. Seventy urban areas(1) met this criterion. For each area population turnover rates were calculated. The rate was defined as those who stated they were not resident in the same dwelling in 1976 as they were a year earlier, expressed as a proportion of the population who could have moved in that period. Thus those less than one year old in 1976 are subtracted from the population of potential movers. In addition, the migration data are compiled only for those who were at their place of usual residence on census night. The result of these adjustments is that the population total for each urban area is somewhat smaller than the official total population. As mobility rates are standardized, this should not prove a problem.

MOBILITY RATES

The significance of the mobility data lies not only in its ability to measure population turnover, but also to analyse the contribution of different forms of movement to the total mobility picture. The Australian Bureau of Statistics migration tabulations enable a very detailed origin destination matrix to be constructed to the level of flows between all pairs of Local Government Areas. While such detail is not employed here, it is possible to examine the components of the total mobility picture, and how this varies between urban areas. Total mobility, measuring all those not in the same dwelling from one period to another, has been subdivided into those who moved either within or between local government areas making up the defined urban area (intra-urban movers), those who moved into the urban area from elsewhere within Australia (inter-regional migrants), and those who were overseas in 1975 (a category including both immigrants, and Australians returning after a stay overseas). A further category, those who had indicated a move, but had not adequately identified the origin, was allocated to the other categories in the proportions in which they represented for each centre. Thus the variation in mobility rates between areas can be examined. Such information is necessary in any attempt to examine both the impact of mobility, and to formulate any explanations of the process. The mobility rates are shown in Table 1, together with the population figure derived from the migration matrices, based on local government areas and an estimate of population change for each area for the 1971-76 inter-censal period. (This latter figure does not conform to the one year period on which the rest of the analysis is based because of the lack of readily available 1975 population figures. It thus can only be used as a general guide to growth rates in the period of interest).

The table establishes a number of facts about mobility in Australian urban areas. Firstly, the figures for annual movement are quite high, with a mean rate of 19.6 per cent of the population having moved at least once during the period 1975-6. The range is considerable however, from a high of 38.3 per cent in Port Hedland, down to 10.3 per cent in Broken Hill, although the standard deviation is relatively small at 4.7 (the rate for Port Hedland is in fact far greater than that for other centres, the next highest being Armidale at 27.5 per cent). The highest rates occur in the resource rich and/or sunshine belt of Western Australia and Queensland where either high labour turnover or immigrants seeking 'the good life' contribute to the figures. The three towns of New South Wales exhibiting high rates reveal a different character. Armidale and Bathurst both have relatively large proportions of tertiary students resident only for their courses and Bathurst has also benefited (at least in the 1975-6 period) from its designation as a growth centre. By contrast Queanbeyan's high rate is due to its proximity to Canberra. Low mobility appears to be common to areas with little in-movement and low levels of growth, or even population loss. Apart from Broken Hill, areas such as Port Pirie, Burnie, Colac and Lthgow experience little turnover (less than 15 per cent) as do all the towns in the La Trobe Valley of Victoria (Moe/Yallourn, Norwell, Traralgon), and the two larger industrial centres of Newcastle and Wollongong.

Although total mobility has a number of components it appears that the variation in total movement is dependent on the degree of in- and out-movement. This has generally been related to an attractiveness function which can in turn be related to the nature of the economic base (particularly demand for labour in the resource industries), or to environmental attractions such as climate. Intra-urban mobility, on the other hand, is less variable, having a standard deviation of only 1.9 around the mean of 9.7 per cent for the 70 centres. The range is from 15.1 per cent of the population in Perth to a low of 5.8 per cent in Gympie. Areas of higher intra-urban mobility are the state capitals, all of which exceed 11 per cent, as well as Cairns, Kalgoorlie, and Townsville. Low intra-urban movement is recorded in Gympie, Broken Hill, Horsham, Morwell and Traralgon. The comparatively low level of variation could well be attributable to the nature of intra-urban mobility which, it has been argued in the literature, is largely a function of changing accommodation needs of households as they move through the life cycle. Variations in intra-urban mobility would then be attributable to variations in the composition of the population in each area (Speare, Goldstein and Frey, 1974).

MOBILITY

TABLE 1 Mobility Rates in Australian Urban Areas 1975-76

	1976 Popn (1)	Popn change 71-76 (2) %	Total mobility %	Intra-urban %	Inter-regional %	Over-seas %	Outmoves (3) %
1. Adelaide SA	884921	6.9	16.4	11.8	3.4	1.2	2.7
2. Albany WA	12445	1.1	23.7	11.7	11.2	0.8	13.5
3. Albury Vic	31504	15.9	22.8	11.3	10.7	0.8	9.9
4. Armidale NSW	17914	8.6	27.5	11.5	14.3	1.8	11.8
5. Ballarat Vic	71957	6.0	16.3	9.9	5.9	0.4	5.2
6. Bathurst NSW	17719	8.1	25.2	9.8	14.4	1.0	9.9
7. Bendigo Vic	58383	10.4	17.6	9.8	7.2	0.6	5.0
8. Brisbane Qld	837647	10.1	18.1	11.5	5.2	1.3	4.6
9. Broken Hill NSW	27115	-7.2	10.3	6.1	3.8	0.3	5.7
10. Bunbury WA	18993	9.5	23.8	11.5	11.5	0.8	10.1
11. Bundaberg Qld	29437	11.5	19.1	8.8	9.8	0.6	10.0
12. Burnie Tas	19095	-2.5	14.3	7.1	6.6	0.6	8.2
13. Cairns Qld	51483	30.9	24.9	12.4	11.3	1.3	9.8
14. Caloundra Qld	14333	50.1	24.6	7.6	16.2	0.9	9.3
15. Canberra ACT	154939	37.5	23.9	10.5	11.0	2.5	7.1
16. Coffs Habour NSW	22645	31.5	23.5	11.1	11.6	0.7	6.8
17. Colac Vic	15612	-1.3	14.2	8.1	5.9	0.2	6.7
18. Darwin NT	27489	18.2	20.9	8.6	11.4	0.8	10.0
19. Devonport Tas	21098	8.2	16.6	8.9	7.2	0.5	7.0
20. Dubbo NSW	19287	12.6	20.9	10.5	9.9	0.3	8.6
21. Geelong Vic	150983	10.9	15.9	10.1	5.0	0.8	3.7
22. Geraldton WA	16833	17.1	27.3	10.9	15.4	1.1	12.6
23. Gladstone Qld	18132	18.0	25.5	11.6	12.8	1.1	11.2
24. Gold Coast Qld	141671	38.8	25.8	10.7	13.6	1.5	7.9
25. Goulburn NSW	20588	0.6	19.2	10.5	8.0	0.6	6.8
26. Grafton NSW	16070	0.8	16.1	7.9	7.7	0.4	8.3
27. Griffith NSW	18918	3.8	18.6	10.9	7.0	0.7	6.0
28. Gympie Qld	10793	0.9	16.0	5.8	9.8	0.4	12.3
29. Hobart Tas	163940	5.4	16.7	11.5	4.1	1.1	3.8
30. Horsham Vic	11315	5.5	15.5	6.5	8.5	0.6	6.8
31. Ipswich Qld	67462	12.4	17.4	7.9	8.7	0.8	8.0
32. Kalgoorlie WA	19290	-7.2	23.0	12.2	10.0	0.8	17.3
33. Launceston Tas	76230	3.9	15.4	9.8	5.0	0.6	4.4
34. Lismore NSW	21342	5.6	18.3	8.3	9.4	0.6	9.1
35. Lithgow NSW	12051	-3.8	13.7	7.3	5.7	0.7	6.7
36. Mackay Qld	44324	13.1	21.8	11.3	9.7	0.8	8.4
37. Maroochy Qld	31771	38.2	24.0	9.6	13.3	1.1	8.8
38. Maryborough Qld	20837	4.6	16.5	7.9	8.1	0.5	7.3
39. Melbourne Vic	2552344	4.0	16.8	13.2	2.1	1.4	2.6
40. Mildura Vic	13766	1.6	20.0	8.5	10.9	0.6	12.3
41. Moe/Yallourn Vic	16911	-8.7	14.1	7.6	6.1	0.4	7.9
42. Morwell Vic	22314	0.9	14.7	6.8	7.2	0.7	7.5
43. Mt Gambier SA	17425	2.7	18.4	9.4	8.5	0.5	10.7
44. Mt Isa Qld	24860	0.1	23.5	11.9	10.7	0.8	15.8
45. Newcastle NSW	356221	3.3	13.9	9.9	3.4	0.6	3.2
46. Nowra NSW	36172	29.6	21.9	10.9	10.2	0.7	8.4
47. Orange NSW	24773	10.0	21.5	11.4	9.7	0.4	8.2
48. Pt Augusta SA	12510	7.1	16.1	8.1	7.8	0.3	10.2
49. Perth WA	784422	14.6	22.1	15.1	5.1	2.0	3.3
50. Pt Headland WA	9771	57.9	38.3	11.8	22.9	3.6	24.4
51. Pt Lincoln SA	9878	6.8	19.6	9.5	9.1	0.9	11.4
52. Pt Macquarie NSW	11921	43.0	22.9	9.6	12.6	0.7	8.1
53. Pt Pirie SA	12196	-6.7	12.4	7.3	4.9	0.2	9.4
54. Queanbeyan NSW	18223	17.7	25.2	8.2	15.7	1.2	15.5
55. Rockhampton Qld	49212	4.0	16.3	8.5	7.2	0.6	7.8
56. Sale Vic	11805	15.6	23.2	8.9	12.9	1.4	13.2
57. Shepparton Vic	40216	6.9	17.2	9.7	7.0	0.5	6.9
58. Sydney NSW	2920925	2.9	17.0	13.3	2.1	1.6	2.5
59. Tamworth NSW	28337	10.8	19.5	10.3	9.0	0.3	7.6
60. Taree NSW	12499	12.4	19.4	8.2	10.7	0.4	8.1
61. Toowoomba Qld	63931	11.6	20.3	10.6	9.0	0.7	7.8
62. Townsville Qld	85311	19.8	26.6	13.1	12.2	1.2	9.2
63. Traralgon Vic	14765	2.9	14.6	6.7	8.1	0.4	7.5
64. Tweed NSW	25996	14.4	18.7	7.9	10.2	0.5	7.0
65. Wagga Wagga NSW	33170	18.8	23.4	11.5	11.2	0.7	10.8
66. Wangaratta Vic	15687	3.4	16.7	7.1	9.3	0.3	8.2
67. Warrnambool Vic	19768	8.1	17.8	8.9	8.1	0.7	7.2
68. Whyalla SA	32446	5.8	15.6	8.8	5.5	1.3	10.2
69. Wodonga Vic	14966	20.3	24.6	8.1	15.9	0.6	10.7
70. Wollongong NSW	206536	6.0	14.9	10.8	3.1	1.1	3.9
Mean		11.1	19.6	9.7	9.1	0.8	8.6
Std Dev.		13.2	4.7	1.9	3.8	0.6	3.7

NSW = New South Wales; Vic = Victoria; SA = South Australia; Qld = Queensland; WA = Western Australia; NT = Northern Territory; ACT = Australian Capital Territory; Tas = Tasmania

(1) 1976 Population is that over one year old in 1976, and at their usual place of residence.
(2) Source: Australian Bureau of Statistics, 1976 Census: Population and Dwellings in Local Government Areas and Urban Centres (Preliminary) (2401.0, 2408.0) (N.B. Where LGA boundaries have changed, the 1971 figures are given for the 1976 boundaries).
(3) Excludes those who moved overseas.

Source: ABS Unpublished Matrix Tapes (Matrix 14).

On the other hand, Moore (1981) has argued that variations in the rates of intra-urban mobility may be explained in part by variations in the local context, particularly after compositional differences have been taken into account. The local context could reflect the particular nature of housing types, vacancy rates and the organisational characteristics of the housing market which may vary from place to place. The impact of of compositional and contextual effects is being investigated elsewhere (Maher, forthcoming, a).

Comparisons of intra-urban mobility rates are more difficult than those of total mobility because of the definition of what is intra-urban. Clearly the area of a metropolis such as Melbourne or Sydney allows quite long moves to be recorded as intra-urban, whereas where the urban boundary is that of one LGA, moves across this are recorded as external moves. At the same time, it could be expected that intra-urban movement would be greater in larger urban areas. With increasing population size comes a greater range of potential locations for both residence and for other types of activity, especially employment. The more highly dispersed and segmented nature of housing and employment sub-markets in metropolitan areas in particular would, other things being equal, result in a higher level of intra-urban movement. The scope to alter one's type of residence, or location of employment is clearly going to be less in Gympie or Horsham than in Melbourne or Sydney.

Differences in the rates on in-movement either from elsewhere in Australia or overseas, tends to repeat the pattern for total mobility. Highest rates are Caloundra, Geraldton, Queanbeyan and Wodonga. The results for the last two are very much a function of their proximity to other larger centres. Low in-movement from elsewhere in Australia is a feature of most of the metropolitan areas (except Perth), partly a function of the large population base, but also reflecting a decline in the ability of the large metropolitan areas to attract population over the past few years (Burnley, 1980).

In-moves originating from overseas show a slightly different pattern. Although only a small proportion come from overseas in any one year (a mean figure of 0.8 per cent) higher proportions are destined for the metropolitan areas, as well as Gold Coast and Port Hedland. Apart from Melbourne and Sydney which traditionally attracted the bulk of overseas migrants, Canberra also has a higher overseas component. This no doubt is due to overseas postings in government positions, public service recruiting, and the highly mobile nature of the university community. In Western Australia, the higher levels for Perth and Port Hedland are presumably related to the demands for skilled labour which have occurred while the Gold Coast reflects the ability of that area to attract retirees not only from within Australia, but also from overseas. Those areas with little in the way of overseas arrivals are the smaller, largely inland towns such as Colac, Dubbo, Tamworth and Wangaratta.

The sixth column in Table 1 describes the extent of out-movement from each area. It is a much more difficult figure to compute as it involves, for any one LGA, scanning the origins of migrants to all other LGAs in Australia. This method, of course, fails to pick up that component of the population who had left for overseas between 1975 and 1976 as they would have answered the census. The areas that had the highest out-movement (for example, Kalgoorlie, Mount Isa and Port Hedland) were clearly areas with high rates of labour turnover and as such tended to feature high in-movement rates as well. The difficult physical environment and the isolation of these towns means that there is a constant flow through of population. At the other end of the scale, lowest out-movement was associated with the large metropolitan areas, and particularly with Adelaide, Melbourne and Sydney. As discussed above, this is partly a function of the number of internal opportunities available to both the job and the house seeker, and partly to the very large population base.

A final comment on Table 1 relates to growth rates. Although the figures are for the 1971-76 period, there is an obvious association between mobility rates (particularly in-migration) and growth. Those areas showing the highest growth particularly in coastal Queensland, also have by far the greatest imbalance of in- and out-migration rates. Those areas losing population generally show up as having an excess of out- over in-movers. Some exceptions do exist however, and act to demonstrate the manner in which population trends can change in a short time. Two examples are Sale and Whyalla, both of which grew in the 1971-76 period, but had an excess of outmovers in the 1975-76 period.

In summary, the Australian urban population is a highly mobile one. An average of almost one person in five in urban areas over 10,000 in population moved residence in the period 1975-76. When examined in more detail however, it is apparent that the overall figure masks considerable variability in total rates of movement and also in the relative contribution of sources of turnover. Before examining the implications of these movement rates however, more detail was sought on the turnover sources themselves, particularly with respect to the nature of their interrelationships and the effect of differential population size and growth rates.

INTERRELATIONSHIPS OF MOBILITY, MIGRATION AND POPULATION

Within recent literature, it has been argued that migration rates are not independent of each other. In particular, it has been suggested that in- and out-movement rates are associated, as are in-movement and intra-urban adjustment. The suggestion derives from the reasoning that the probability of movement for an individual household is a function of the length of time since the last move. Thus notions of cumulative inertia (Cordey-Hayes and Gleave, 1974) or the similar 'duration of stay' effect (Clark, 1981) indicate that the longer an individual household stays in a particular location, the lower the probability of subsequent movement. This is exhibited in the aggregate as a relatively high correlation between on the one hand in-movement and out-movement from an urban area, and on the other by an association between in-movement and intra-urban (Adams, Caruso, Norstrand and Palm, 1973; Goodman, forthcoming; Clark and Everaers, 1981).

To explore these relationships in the context of the nature and rates of turnover in Australia, a correlation matrix was derived, analysing the strength of association between each of the mobility sources, and of mobility with size of urban area and growth rates. The variable size was included in order to examine in more detail the superficial impression that intra-urban mobility in particular was a function of city size. The results are shown in Table 2.

TABLE 2 Simple correlations of population size and turnover rates

	Total Popn	Growth 1971-76	Intra-urban mobil-ity	Inter-regional moves	Over-seas	Out-migration
Total Population	1.0	-0.09	0.43	-0.43	0.30	-0.42
Growth 1971-76		1.0	0.26	0.68	0.56	0.24
Intra-urban mobility			1.0	0.04	0.53	-0.03
Interregional moves				1.0	0.36	0.77
Overseas moves					1.0	0.29
Out-migration						1.0

In fact, over the set of centres ranging in size from ten thousand to three million, the correlation between population size and the various types of mobility is relatively small. There is a positive association between intra-urban mobility and total population, although the correlation coefficient is only is that of immigration from overseas, although this also is small, again indicating that while overseas migrants tend to locate in the larger centres, a number of other factors are important.

The highest correlation shown in the table is between rates of in-movement (inter-regional migration), and rates of out-movement (0.77),a finding which accords with the notion that propensity to migrate decreases as length of residence increases. The association of in- and out-movement is thus explained by reference to the fact that more recent migrants to an area have a greater probability of moving than those who have established a strong network of social and economic ties through a longer period of residence in the same area. Thus areas with the highest rates of turnover are those undergoing substantial rates of both in- and out-migration.

Such an analysis, while demonstrating that some urban areas have much more mobile populations than others, does not explain why this should be so. Clearly high rates of turnover are related to a large number of opportunities being created, opportunities in housing and employment in particular, but also in education and training. Growing areas tend to create job and housing opportunities that act to set off a series of linked moves described as vacancy chains (White, 1970; Maher, 1979). In other areas, a high turnover of population may reflect an annual process (such as tertiary students in Armidale), a seasonal process, or a process reflecting economic booms or depressions.

The 'duration of stay effect' has also been seen to be important in intra-urban mobility (Clark, 1981). Following the concept of cumulative inertia, one might expect a relationship between immigration and intra-urban mobility. Some evidence for this expectation can be found in the literature. For example Adams _et al_. (1973) argued that in-migrants to an urban area frequently move again after a relatively short time in order to adjust their housing, or more likely to move to a location more in tune with their needs. This is done as the household is activity patterns became established and greater knowledge of the area is acquired. They found evidence of this in a sample of new residents in Minneapolis, Minnesota, where 50 per cent of the sample households moved again within a year. Goodman (forthcoming) also found a correlation between in-migration and intra-urban movement, which he explained with reference to an expanding housing stock which created the vacancy chain effect.

However, results in this study do not bear out these findings. A correlation of only 0.04 was found between inter-regional migration and intra-urban mobility, although there was a positive relationship of 0.53 between overseas immigration and intra-urban moves. Why such difference should exist between the United States and Australia is difficult to assess, though part of the discepancy may be attributable to different research methodologies. The work of Adams _et al_. was based on a survey of individual behaviour over a period of time. Aggregate rates over a range of urban areas are not available in the United States. The present study, on the other hand, examines aggregate rates and the results may not be comparable.

Interestingly, in a study of 40 Dutch cities utilizing a similar methodology to that employed here (Clark and Everaers, 1981), there was also little correlation found between in-movement and intra-urban mobility. This was explained with reference to the housing system in Holland, where complex sets of eligibility rules, housing shortages, and the difficulty of obtaining finance, were all seen to make it likely that further adjustments would be made after the initial location. While these reasons may well be valid, it is also likely that, when taking account of a wide range of different urban contexts, differences may well emerge that would not be considered when examining the behaviour of a small sample in just one area.

While the correlation between in-migration and intra-urban mobility is low, graphing the relationship and being able to identify each point can be of assistance in interpreting the relationship over the entire range of urban areas (Figure 1. Each observation can be identified by the numbering in Table 1). Two points emerge from the plot. Firstly, there is an overall impression of a positive relationship between in-migration and intra-urban mobility. However, there are two groups of observations which do not conform to this pattern. One set has high rates of intra-urban movement, but low rates of in-movement. These are the metropolitan centres, and most noticeably Perth (number 49 on the diagram), Sydney (58) and Melbourne (39). On the other side of the diagram, the other set has relatively high rates of in-movement, but low rates of intra-urban moves. The centres in question are Caloundra (14) (a resort area to the north of Brisbane), Queanbeyan (54) (closely associated with Canberra) and Wodonga (69) (a sister city with Albury in New South Wales). These three could easily be merged with the towns they are close to and much of the in-movement would then be redefined as within-area moves.

TABLE 1 Correlations between intra-urban mobility and inter-regional mobility.

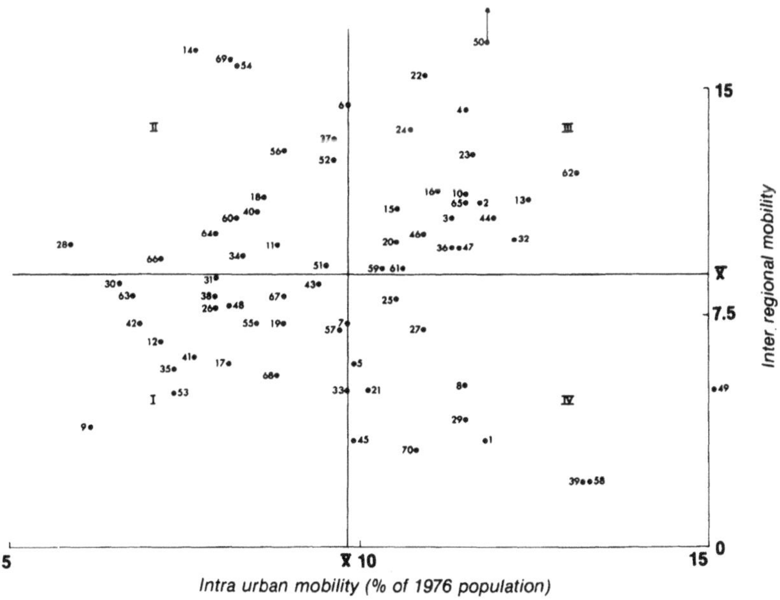

MOBILITY

A second way in which the diagram can be used to interpret variations in mobility across the urban areas is to divide it into four quadrants. The divisions are drawn at the mean value of migration and intra-urban mobility on the respective axes. These quadrants then can be described in terms of the two mobility sources. Quadrant 1 for example in the lower left of the diagram, contains towns low in both intra-urban and in-movement relative to the mean figure; quadrant II contains those low on intra-urban rates but high on in-movement; quadrant III has those centres with high scores on both axes, and quadrant IV includes those high on intra-urban movement, but low on in-movers. By identifying towns with characteristics in common, a rudimentary typology of towns based on mobility behaviour can be identified. This should serve as an initial means of exploring why such variations in movement behaviour occur.

The first quadrant, defined by low rates of movement on both axes, contains centres where there is both little attractiveness for in-movement, and few internal opportunities for intra-urban movement. This situation typifies many of the small to mid-sized towns of the southern states. Port Pirie (53), Whyalla (68) and Port Augusta (48) in South Australia. Moe/Yallourn (41), Colac (17) and Morwell (42) in Victoria, Burnie (12) and Devonport (19) in Tasmania, and Broken Hill (9), Lithgow (35) and Grafton (26) in New South Wales. These are predominantly declining or slow growth areas. Significantly no Western Australian and only two Queensland towns appear in this quadrant.

Quadrant III on the other hand represents those areas with high rates of both intra- and in-movement. This quadrant is made up almost entirely of towns from the three states of Queensland, Western Australia and New South Wales, the only exception being Canberra (15). Apart fom Port Hedland (50), the Western Australian towns of Geraldton (22), Bunbury (10), Albany (2) and Kalgoorlie (32) are all contained in this sector of the diagram, as are Gold Coast (24), Gladstone (23), Townsville (62), Cairns (13), Mt Isa (44) and Mackay(36) from Queensland and Armidale (4), Coffs Habour (16), Wagga Wagga (65), Albury (3), Nowra (46), Orange (47), Dubbo (20) and Tamworth (59) from New South Wales. Many of these are the booming mining or resource towns, whilst others are the resort and retirement centres on the New South Wales and Queensland coast. It would appear that the differences in these two quadrants represent some intangible element of attractiveness, either in economic or environmental terms. Were growth rates to be included as a means of discrimination between the centres the differences between these two groups would likely become even more marked.

The second quadrant where there are high rates of in-migration but little internal redistribution, is a mixed lot. Most of the centres however, are relatively small areas such that the potential for improving housing or acessibility by internal movement is not great. Quadrant IV on the other hand contains those centres where the greatest range of opportunities exists in terms of choice of type and location of housing. Thus internal redistribution rates here greatly exceed the rates of in-movement, which in the metropolitan areas has slowed markedly in recent years (Rowland, 1979).

OUTCOMES AND INTERPRETATIONS

The analysis has pointed up the highly complex nature, and variable impact of, mobility in Australian urban areas. It now remains to discuss some of the issues which such a pattern raises, and to canvass some of the possible explanations for the results. What has been described here is a set of aggregate outcomes of a process acting continually to redistribute population within and between urban areas. The figures represent the outcome at a particular point in time (the year 1975-76), and it is possible that the observed patterns are partly attributable to conditions existing at that time. However, many of the trends do confirm patterns existing since at least the 1966-71 period, and many of the outcomes of the population mobility have implications which are of a long-term nature.

The outcome of the aggregate patterns has been to point up the differential nature of the mobility process at three particular scales, national, regional and local. The implications at each of these levels tends to vary considerably. The most obvious outcome of migration at any of the three scales is that of a differential growth rate for population. As shown by Table 1, high growth rates are frequently associated with high levels of in-movement, although this effect is obviously modified by the degree ot out-movement occurring at the same time. Even if in- and out-migration are in balance so that growth is not occurring, the second possible outcome of mobility is that the characteristics of the population may undergo a change. This is particularly the case when the mobile population is of a specific age group. A third outcome of mobility relates to the change in the nature of opportunities facing the migrant, and especially the fact of an improvement in economic or social conditions, as a result of the move.

At the national level, it has been recognized for some time (Rowland, 1979; McKay and Whitelaw, 1978) that there is a tendency for population to move away from the most populous states (Victoria and New South Wales), towards Western Australia and Queensland. While the present analysis is not designed to identify interstate migration as such, the migration figures show that the attractiveness of many towns and cities in Queensland and Western Australia are higher than most of those for the other states. High in-migration and rapid growth are noticeable in the towns of coastal Queensland - high in- and out-migration is characteristic of the resource towns of both Western Australia and Queensland; and Perth and Brisbane as the metropolitan centres of these states exhibit higher rates of growth and mobility than do those of the other states.

At a regional level, the picture is more complicated. Even those states experiencing the greatest overall growth show marked differences in the performance of individual towns. In Queensland for example, the urban areas of Gympie, Mount Isa, and Bundaberg all show a rate of out-migration to elsewhere in Australia as greater than that of in-migration. (The overseas component of in-movement is ignored in this comparison because it is impossible to derive the corresponding component of out-movement). In Western Australia, the same situation exists for Albany and Kalgoorlie.

The other states show even greater variability between towns, indicating that population redistribution is occurring within, as well as between regions. New South Wales and Tasmania have a majority of their towns showing an excess of in- over out-migration. Victorian towns are balanced relatively evenly between gainers and losers, and South Australia is losing population through a net migration outward from each of the towns except Adelaide. In South Australia then, the trend toward metropolitan domination is continuing. In New South Wales and Victoria however, the major cities of Sydney and Melbourne show a loss, as do some of the relatively small towns, for example, Broken Hill, Grafton, Lithgow, Colac, and the areas in the La Trobe Valley of Victoria. Gains on the other hand, are being made by the larger non-metropolitan centres such as Newcastle, Albury, Wagga Wagga, Goulburn, Ballarat, Bendigo and Geelong.

At the local level, population redistribution is represented by intra-urban mobility. The metropolitan areas have the greatest number of such moves while mobility to or from these areas is on the whole, quite low. This emphasises the critical importance of immigration to areas such as Sydney and Melbourne. Without migrants originating overseas, the population leaving these two areas would have exceeded that coming in, even given the underestimate of the out-movers through not being able to enumerate those who have left for an overseas destination. If these trends continue, the future growth of both Australia's major cities will therefore be dependent on decisions made at a level far removed from a concern with cities, that of the immigration programme. Within the larger metropolitan areas rates of intra-urban movement are uniformly higher, reflecting the significance of population redistribution at another scale, that of the local area level within cities. Here intra-urban movements are predominantly a reflection of the continuing process of suburbanization (Maher, forthcoming, b) and although the measurements are quantitatively complex the most significant of these is the redistribution of population outward. Even in an era of slow or no-growth, the pull of the suburbs remains strong. The implications of movement of this nature and magnitude are great for the provision of local services. As a population distribution alters, areas which were once adequately served become grossly over-served, while others are in great need of even minimal support. In the intra-urban case, one of the best examples is that of education planning where not only quantitative, but also qualitative movements have implications for the amount and type of educational services required.

Overall then, the aggregate patterns of migration indicate the nature of population redistribution occurring within the urban system. This includes the growth of Queensland and Western Australia at the expense of most of the other states; a growth of the smaller metropolitan areas and larger non-metropolitan areas, and the relative decline of the two major metropolitan areas and some of the smaller, older urban settlements; and the continued movement of population within the larger urban areas, particularly outward toward the periphery.

The outcomes of population mobility, and the issues they raise, are of vital importance in the planning and implementation of policies and programmes devised to cater for the needs of a changing population. Equally important is the ability to understand why these various movements are occurring. In an earlier section of this paper, some of the general explanations which have been offered were discussed. In particular, a distinction was made between the two major scales of migration (total and partial displacement), and the two major models of explanation (macro- and micro-scale).

Explanations of the inter-regional scale of movement have concentrated on economic factors (high unemployment, low job choice at source; greater employment opportunities at destination). Increasingly, a second type of explanation, relating to environmental factors (climate, beaches and other physically appealing features) is becoming important in encompassing certain types of movement. However reference to broader structural arguments is also necessary. The increased or decreased economic performance by particular regions which is seen to be a cause of many total displacement moves is itself not independent of broader economic, social and political factors. Many of the regional differences in growth and mobility are not only due to varying endowment in physical resources but also to government economic policy at state and national levels, and to national and international economic conditions. Likewise the move to the Sunbelt can be seen as a response to similar forces, and not just to a search for sunshine. The growing ability of retirees to afford to move and seek out investment as well as climatic opportunities must be seen in the light of the wider availability and growing size of lump sum superannuation payments. At the other end of the economic scale, the unemployed may well find it easier and cheaper to live in a climate which places less demand on expensive housing, heating bills, or even the necessity to maintain a varied range of clothing. Thus migration patterns which on the surface may have a relatively simple explanation, in fact are much more a function of structural forces.

The level of complexity inherent in the mobility process has been more readily accepted at the level of intra-urban moves, but here also explanations have tended to be somewhat simplistic. The major thrust in accounting for movement within urban areas has drawn on the relationship between household attributes and housing needs. The majority of intra-urban moves have been seen to be a response to the changing housing needs of a household as it progresses through the life cycle. Evidence is provided by the fact that most moves are short, generally within the same local area (and hence the partial displacement categorization). Increasingly however, the emphasis has shifted away from the immediate needs of the household, towards a consideration of the organization of the system of providing and allocating dwelling units. Much of the difference in mobility between the United States and Britain for example, has been attributed to the much greater role of the public sector in the latter. The housing system in Australia is not greatly different to that in the United States, at least in its largely private market orientation. This adds to the extent of intra-urban mobility in a number of ways. First, households are freer to make relatively small adjustments to their housing needs. While there are some impediments to movement, such as leases and bonds in the rental sector, and certain transaction costs in transferring ownership, there are no intricate rules of qualification which impede mobility in the public sector. Second, because private ownership is seen not only as a secure means of tenure, but also a means of investment, some households may be tempted to move solely to capitalise on an increased level of equity and thus to extend their consumption of housing services, whether these are actually needed or not. Finally, because in a housing system such as that existing in Australia, considerable benefits attach to the ownership of real proerty (Paterson, 1974), the eventual goal of most families is to achieve the status of a home-owner or purchaser. Thus many moves are made solely to effect this. It may not represent the ultimate in the dwelling aspirations of the household, but the assumption is that once ownership has been achieved then the household has a better chance of being able to eventually gain the house of their dreams through acruing capital gains in one or a succession of properties.

Factors broader than the organization of the housing system itself are also very important when considering the rates of intra-urban mobility. Changing social patterns, such as the rising proportion of single person or non-family households, may mean a higher proportion of households are not interested in owning property. At the same time, a considerable rise in the rate of marital dissolution over the past ten years is affecting both the rate of intra-urban mobility and the nature of household formation. Economic factors such as rising interest rates must also have an impact on mobility patterns. An increasing proportion of households will be unable to obtain housing finance, and thus are likely to remain in the more mobile rental sector. However for those already owning, rising interest rates may force the postponement of plans to upgrade a dwelling.

Clearly structural factors are of great significance in the mobility process and more work needs to be done in this area. However, evidence from this research as well as elsewhere, demonstrates that even within the same overall context of social, economic and political factors at the intra-national level, considerable variation in the extent of mobility exists (particularly intra-area mobility). A fruitful avenue of research lies in the identification of why within-area mobility should vary so widely within a broadly similar housing system. The little work that has been done in this area so far has identified two major explanations. The first is what can be termed the compositional effect (Speare, Goldstein and Frey, 1974) which suggests that population groups with certain characteristics are much more likely to move than others and that some urban areas have higher proportions of high mobility (or low mobility) populations than others. Those characteristics which identify the most mobile population are life cycle (young adult), marital status (unmarried, or recently married but childless), and dwelling tenure (renter). For example, the high mobility of an area such as Armidale is clearly a compositional effect.

The second approach to examine variability in mobility rates between areas is what has recently been termed the contextual effect (Moore, 1981). Using a limited range of evidence from the United States' Housing Allowance experiment, Moore argued that at the local level, contexts tend to vary quite markedly. The local context can be described in a number of ways, but generally includes measures of population size and growth rates, economic and workforce indicators, and measurements of housing market activity such as vacancy rates, age and type of dwellings, and rates of new construction.

The observed differences in intra-urban mobility levels between the urban areas examined in this paper can therefore be related to both compositional and contextual effects and work on this is presently underway (Maher, forthcoming, a). From the evidence of considerably higher rates of intra-metropolitan movement noted earlier, it is clear that the contextual effects operating in such centres are much more appropriate to movement than those in the larger non-metropolitan areas or in small towns.

Thus mobility is a highly complex process and no simple theories are likely to be derived to explain it (Moore and Harris, 1979). The process is a part of the operation of labour and housing markets; it is at once an expression of individual needs or desires, local contextual and compositional variations, and the organisation of social and economic factors at the structural level; and it occasions improvements or changes, not only in the circumstances of individual households, but in the aggregate, and affects the general redistribution of population and population characteristics in time and space. As was stated at the outset, migration and mobility are basic to the dynamic of an urban system, a fact which justifies its study. The complexity of the process makes a better understanding of it difficult, yet the necessity remains. This paper has presented some elementary empirical trends in population movement in Australian urban areas, and has suggested how studies of this might be extended. Two broad thrusts are needed. On the one hand to develop the conceptual apparatus for understanding the complexity of the process and on the other, in Australia in particular, we need to know more about the nature and rates of population movement, particularly in a temporal frame. It is hoped that this paper may have assisted in a clarification of each of these.

NOTE

1. The definition of urban area used in this study is not the same as that of urban place utilized by the Australian Bureau of Statistics. Urban places are defined on the basis of the extent of the built up area defined by density of population in collectors' districts. However, the mobility data are tabulated by Local Government Area (LGA), the boundaries of which do not necessarily coincide with those of the urban place. In order to utilize the mobility data, it was necessary to approximate the urban area boundary with that of LGA boundaries. This has the effect in some cases of including as urban, LGAs which are only partially urban in character. This occurs particularly at the periphery of large urban agglomerations. The effect is to overestimate the population of some centres by including some non-urban component. In fact, this is frequently done when metropolitan statistical divisions are used in studies of major metropolitan areas. Metropolitan statistical division boundaries take in considerably more than the contigious urban population.

   The variation from urban to LGA population noted in this study is on the whole not great. In the majority of cases the approximated population is little different from that of the urban place. Differences arise in a few cases where settlement is widespread, frquently in a semi-urban situation. This is most marked in the resort areas of northern New South Wales and Queensland, LGAs such as Coffs Harbour, Tweed, Gold Coast, Maroochy and Cairns. Despite the overestimation of population in these areas however, it can be argued that in situations of this nature, it makes more sense to include the whole of the population because in these areas there is not the great distinction between urban and rural populations that there is elsewhere.

REFERENCES

Adams, J., Caruso, D., Norstrand, E., and Palm, R., (1973) 'Commentary: intra-urban migration', Annals, Association of American Geographers, 63:152-55.

Bassett, K. and Short, J.R., (1980) Housing and Residential Structure: Alternative Approaches. London: Routledge and Kegan Paul.

Bell, C. (1977) 'On housing classes', Australian and New Zealand Journal of Sociology, 13:36-40.

Brown, L.A. and Moore, E.G., (1970) 'The intra-urban migration process: a perspective', Geografiska Annaler, 52B:1-13.

Burnley, I.H. (1980) The Australian Urban System. Melbourne, Longman-Cheshire.

Burnley, I.H., Pryor, R.J. and Rowland, D.T., (1980) Mobility and Community Change in Australia. Brisbane, Queensland University Press.

Castells, M., (1977) The Urban Question. London, Edward Arnold.

Cities Commission (1974) Studies of Australian Internal Migration, 1966-71 Occasional Paper No. 2. Canberra, Cities Commission.

Clark, W.A.V., (1976) 'Migration in Milwaulkee', Economic Geography, 52:48-60.

Clark, W.A.V., (1981) 'Some observations on the structure of recent research on migration and mobility', Geografische en Planologische Notities No. 10, Amsterdam, Free University.

Clark, W.A.V. and Everaers, P., (1981) 'Public policy and residential mobility in Dutch cities',Tijdschrift voor Economische en Sociale Geografie, 72, 6:322-333.

Clark, W.A.V. and Moore, E.G., (1978) Population Mobility and Residential Change, Studies in Geography No. 25, Evanston, Northwestern University.

Clark, W.A.V. and Moore, E.G. eds.,(1980) Residential Mobility and Public Policy, Urban Affairs Annual Reviews, 19, Beverley Hills, Sage.

Clark, W.A.V. and Moore, E.G., (forthcoming) 'Residential mobility and public programs: current gaps between theory and practice'. Forthcoming in special volume of Social Issues.

Cordey-Hayes, M. and Gleave, D. (1974) 'Migration movements and differential growth of city regions in England and Wales', Papers and Proceedings of the Regional Science Association, 3, 3:99-123.

de Castro Lopo, L., (1974) Internal Migration in Western Australia: A Search for Order in Migration Patterns. Unpublished Ph.D. thesis Newcastle, University of Newcastle, New South Wales.

Gale, S., (1978) 'Remarks on information needs for the study of geographic mobility', in Clark, W.A.V. and Moore, E.G., eds., Population Mobility and Residential Change, Studies in Geography, No. 25, Evanston, Northwestern University.

Gale, S., (1980) 'Alternative Agenda in academia and public policy', in Clark, W.A.V. and Moore, E.G., eds., Residential Mobility and Public Policy, Urban Affairs Annual Reviews, 19:34-38, Beverley Hills, Sage.

Goodman, J., Migration, mobility and information, in Clark, W.A.V. ed., Mobility, Migration and the Search for Housing. (in preparation).

Gray, F. and Boddy, M., (1979) 'The origins and use of theory in urban geography: Household mobility and filtering', Geoforum 10:117-127.

Hanushek, E.A. and Quigley, J.M., (1978) 'Housing market disequilibrium and residential mobility', in Clark, W.A.V.

and Moore, E.G., eds., Population Mobility and Residential Change, Studies in Geography, No. 25, Evanston, Northwestern University.
Harloe, M., (1977) Captive Cities. London, Wiley.
Harris, R. and Moore, E.G., (1980) 'An historical approach to mobility research', Professional Geographer, 32:22-29.
Harvey, D., (1975) 'The political economy of urbanization in advanced capitalist countries: the case of the U.S.', in Gappert, G., and Rose, H., eds., The Social Economy of Cities, Urban Affairs Annual Review, 9, Beverley Hills, Sage.
Humphries, J.S., (1973) Intra-urban Migration and Residential Structure. Monash Publications in Geography, No. 6, Melbourne, Monash University.
Kemeny, J., (1977) 'A political sociology of home ownership in Australia', Australian and New Zealand Journal of Sociology, 31:47-52.
Kendig, H., (1981) Buying and Renting. Canberra, Australian Institute of Urban Studies.
Kilmartin, L. and Thorns, D., (1978) Cities Unlimited, Sydney, Allen and Unwin.
McKay, J. and Whitelaw, J.S., (1977) Internal migration and the Australian urban system. Progress in Planning, 10:1-83.
Maher, C.A., (1979) Private Housing Construction and Household Turnover: a Study of Vacancy Chains in Melbourne's Housing Market, Monash Publications in Geography, No. 22., Melbourne, Monash University.
Maher, C.A., (1982) Australian Cities in Transition. Melbourne, Shillington House.
Maher, C.A. (forthcoming, a) Household Mobility and Residential Change in Australian Cities, Census Monograph Series, Canberra, Australain Bureau of Statistics.
Maher, C.A., (forthcoming, b) 'Population turnover and spatial change in Melbourne, Australia', Urban Geography.
Moore, E.G., (1969) 'The structure of intra-urban movement rates: an ecological model', Urban Studies, 6:17-33.
Moore, E.G., (1981) 'Inter-city contrasts in residential mobility: some implications for public policy', Abstracts: 23 Association of American Geographers Conference, Los Angeles.
Moore, E.G. and Harris, R., (1979) 'Residential mobility and public policy', Geographical Analysis, 11:175-83.
Newton, P., (1977) 'Patterns of population movement in an Australian city', Division of Buiding Research, Technical Paper No. 19., Melbourne, C.S.I.R.O.
New Zealand, Department of Statistics (1976) New Zealand Census of Population and Dwellings, 1971. Vol. 11, Internal Migration, Wellington.
Paterson, J. (1975) 'Home owning, home renting and income redistribution', Australian Quarterly, 4, 4:28-36.
Quigley, J.M. and Weinberg, D.H., (1977) 'Intra-urban residential mobility: a review and synthesis', International Regional Science Review, 2:41-66.
Roseman, C.C., (1971) 'Migration as a spatial and temporal

process', Annals, Association of American Geographers, 61: 589-598.

Rossi, P.H., (1980) Why Families Move. 2nd Edition. Beverley Hills, Sage.

Rowland, D.T., (1979) Internal Migration in Australia. Census Monograph Series, Canberra, Australian Bureau of Statistics.

Rowland, D.T., and Mikkelson, L., (1979) 'Australian internal migration: source materials and trends', in Price, C.A., ed., Australian Immigration: a bibliography and digest. No. 4, Canberra: Australian National University.

Short, J.R., (1976) 'Social-systems and spatial patterns', Antipode, 8:77-83.

Short, J.R., (1978) 'Residential mobility', Progress in Human Geography, 2:419-447.

Simmons, J.W., (1968) 'Changing residence patterns in the city: a review of intra-urban mobility', Geographical Review, 58:622-651.

Simmons, J.W., (1974) Patterns of Residential Movement in Metropolitan Toronto. Department of Geography, Research publications, Toronto, University of Toronto.

Speare, A., Goldstein, S., and Frey, W.H., (1974) Residential Mobility, Migration and Metropolitan Change. Cambridge, Mass, Ballinger.

White, H.C. (1970) Chains of Opportunity, Cambridge, Mass, Harvard University Press.

Whitelaw, J.S. and Robinson, S., (1972) 'A test for directional bias in intra-urban migration', New Zealand Geographer, 28:181-93.

6 The city, state and resource development in Western Australia

Elizabeth Harman

There is a widespread piece of conventional wisdom which says that Perth is doing well from contemporary mining booms in Western Australia. This paper is a critique of that assumption and can be read at two levels. First, it can be read for the empirical evidence provided on the impact of resource development on Perth. In writing the paper however, I have been equally concerned with a problem of theory - namely, the uncritical application of a centre-periphery notion (whether orthodox or class-based) to explain the relationship between Perth and its regions. Since the richest of these ideas are those derived from world-system and dependency theories, the analysis can be read at a second level which primarily addresses the extent to which Perth can be understood as one link in a chain of dependency relations.(1)

Among social scientists, the assumption that Perth is a 'winner' is quite probably derived from models which describe the metropolis as dominant in a regional system. Although the specific details vary, some version of the centre-periphery relationship in which cities enjoy advantages over their hinterlands are explicit in central place theory, urban base models, the notion of cumulative causation, Canadian staples theories and the dependency theories associated with explanations of third world development. This dominance is assumed to be greatly strengthened where a city is clearly primate, as Perth is in Western Australia. The size ratio between Perth and Bunbury, the second largest centre, is now 1 : 43. It is equally well recognised, however, that while Perth may enjoy an obvious superiority at the state and regional level, it occupies a peripheral position in national and international economies. Nationally, Perth is subordinate to the financial and manufacturing bases of Melbourne and Sydney. The eastern metropolises are, in turn, seen as subordinate to Tokyo, London and New York in the international circuits of capital. The precise terms in which the international relations of dominance-dependence are described depends on the historical period and theoretical perspective. Descriptions range from theories of colonisation which subordinate Australia to England to more recent notions such as the Pacific Rim thesis which gives Japan a dominant role, while Mullins (1980), has combined the two in his discussion of 'old' and 'new' dependency relations.

Given this pattern, it is very tempting to assume that the impact of post war mining developments in Western Australia can be analysed using the notion of a chain or hierachy of power relations linking successively stronger classes within larger geographic regions. The notion of a hierachy of dependence relations obviously has its origins in Gunder Frank's description of the patterns of Latin American development as a chain of metropolis-satellite relations in which class relations coincide with regional, national and international geographies (Frank, 1969). On the surface, an analogous model seems to provide a simple and elegant expression of the observed patterns of urban and regional relationships in Western Australia and Australia

An interest in the urban and regional implications of Australian resource development has only emerged in recent years as it has become evident that the combined effects of mining and energy investment on the one hand, and deindustrialisation on the other, may be reshaping the economy and geography of this country.

It seems an opportune time, then, to consider the applicability of a model of urban and regional relations based on the centre-periphery notions which dominate so much of the literature. This paper presents an examination of relationships which have evolved between Perth and its resource rich hinterland in the Pilbara, Kimberley and Darling Ranges in the context of mineral development since the 1960's. The basic thesis is that the metropolitan dominance relations which are depicted both in the centre-periphery notion and its elaboration as a chain or hierachy are not appropriate for understanding the tendencies which are affecting Perth. The major problem is at the State level. There is no real dispute with the view that Perth is subordinate to the eastern metropolitan seaboard, although not all communications linking Western Australia with the rest of the world necessarily go via the east as the idea of a linked chain would suggest.

The central argument is that the term 'dominance' gives a misleading impression of the role played by Perth in resource development and the benefits enjoyed by metropolitan residents. When dominance is defined solely in demographic terms then Perth is certainly dominant. It is not only primate - it is still capturing the greatest share of the State's population growth and annually has increased its proportion of the total population. 70.5 per cent of people lived in the city in 1977; 70.8 per cent in 1978; 71.1 per cent in 1979. But dominance is much more than numbers of people. It assumes a bias towards the metropolis in economic, social and political terms - for example, as the centre of innovation; technology and industry; in unequal exchange relations; in its ability to extract surplus from the regions; and as the focus of class and political power. Historically, Perth may have enjoyed such status in its role as port and service centre for the agricultural sector which dominated the economy prior to 1950. Even then, however, it more likely corresponded to some form of dependent urbanism acting as a channel for the export of surplus in the sense outlined by Harvey (1973 : 232). Indeed, Perth has never had a well-defined indigenous ruling class of sufficient import this century to control its own development (Bolton, 1972) and the locus of economic power has been elsewhere. The economy has been traditionally oriented to the export of raw materials (originally agricultural and now the export of minerals). The small secondary sector which existed prior to 1950 was primarily based on food processing. With the advent of the modern phase of mining in the 1960s, Perth's position as 'dominant' over other regions has not strengthened despite the impression given by the growth in metropolitan population. It has, on the contrary, weakened.

The argument is considered in three sections. The first section reviews the logic which has generally supported the notion of metropolitan dominance. The main argument is then pursued on two grounds. One relating to the nature of state intervention in resource development in Western Australia; the other to the logic of capitalist enterprise in mining and auxiliary activities in the State. The discussion of government policy serves to demonstrate how the activities of the State may interfere with any tendency for Perth to capture special advantages from regionally-based mining. The government has sought to attract investment by (among other things) financing infra-structure associated with major resource projects. There has been a consequent shift of public funds: firstly, from welfare to productive economic activity; secondly, from service-dependent populations to private capital; and thirdly, from city to region.

Part of the government's rationale has been that resource-related regional expenditures will return economic activity, investment, jobs and income to the metropolitan area as a result of multipliers stimulated by the mining activity. The irony is that the operation of the major enterprises in the mining sector, in fact, impedes this process. The mining transnationals make their major decisions and relocate profits out of state. More important, with respect to State policy, the major contractors and purchase suppliers are <u>not</u> local. As a result the Perth-based manufacturing sector has not been growing as strongly as the government has suggested (Harman, E. 1981a).

Perth has, nonetheless, had a significant role to play. In response to the needs of companies based in the east or overseas the metropolitan area has become a clearing house - a sort of commercial siphon or brokerage for business transactions which involve on the one hand an inflow of investment, and on the other, an outflow of commodities, contracts, purchase orders and profits directly or indirectly derived from mining in the mineral-rich outer areas of Western Australia. Both private firms and public agencies are engaged in providing export-led services and together they account for a significant growth in tertiary, or more properly, quaternary employment over the last two decades.

In this service capacity, Perth has no special advantage over its hinterland. It simply fulfills a different function for mining capital. Nor are activities in the resource areas necessarily dependent on decisions taken in Perth. City and hinterland perform specialised functional roles, but are equally dependent on decisions taken out of state and internationally by the brokers controlling activities associated with mining. In limited ways, the city does appear to be dominant (as in its hold on the population). In other ways, the regions have the upper hand (for example, in the balance of political power). An uncritical application of a centre-periphery model at the level of the State ignores these tendencies and countertendencies.

Metropolitan Dominance and Dependency Theories

There are three distinct bodies of theory in the English language literature referring to uneven spatial development which make a case for metropolitan dominance. These are work in regional economics, staples theory, and theories of imperialism. There is little need to dwell on the logic inherent in the orthodox regional models. All of them suggest that metropolitan dominance is a by-product of the logic of economic processes. The most useful of them in the context of Western Australia is the notion of cumulative causation associated with the writing of Myrdal (1957), Pred (1966), and more recently, of Holland (1976). By this view, the city gains an initial superiority on the basis of some locational, historical or serendipitous advantage. The initial advantage is then strengthened as a result of economies of scale and negative backwash effects, or in the case of Pred's argument, on the basis of a metropolitan bias inherent in the process controlling innovation in the productive sphere, diffusion, decision-making and multiplier effects. Essentially, the metropolis establishes an unbeatable comparative advantage. The growth of Perth in response to resource development has been explained in terms of economic logic both by politicians and academics (for example Stilwell, 1974).

Staples theory is a label given to a body of literature which refers specifically to economic growth in Canada on the basis of staples. A staple is a raw material (like wheat, iron-ore or forest products) which is exported in an unprocessed form. There are a wide range of variants of the theory which differ in terms of whether they are optimistic or pessimistic about the prospects of achieving modernisation or economic independence (autarchy) on the basis of staples export, and whether or not they are based on some form of class analysis. The work has a long history dating from the 1930's and is experiencing a contemporary revival with modern staples theory being influenced by both marxism and dependency theory.(2) Staples theory is important in that it evolved to meet the need to understand a staples-based advanced capitalist economy and hence, it's application to Australia is much less incongruous that the use of dependency models developed to explain conditions in the Third World.

Metropolitan dominance is a central element in staples theory. It was used by a group of early writers (labelled 'the hinterlanders' by Clements and Drache, 1978) to explain the pattern of internal dependency whereby Central Ontario, the manufacturing heartland of Canada, dominates the dependent, staples-based Western provinces. The relationship between Western Australia and the eastern seaboard is obviously analagous. In writing of the regional implications of the Canadian wheat economy and prairie farming in the 1930's, 40's and 50's, writers like Fowke (1957) and Buckley (1955) saw metropolitan dominance not simply as an outcome of 'neutral economic forces', but in political terms. They identified the root of metropolitan dominance in terms of actions of the state, notably the Canadian national policy and pricing system. They recognised that class conflict underlay the uneven pattern of spatial development, placing western farmers as rural independent commodity producers in direct conflict with eastern, metropolitan capitalists controlling the banks and industry.

Contemporary writers following the Canadian tradition have used the centre-periphery notion at both a larger and smaller geographical scale than did the hinterlanders. The history of colonisation and theories of imperialism have been invoked to describe and analyse Canada's dependent and semi-peripheral role with respect first to Britain and now to American capital and the American economy. Inside Canada, at the provincial level, the major cities are seen as creating their own hinterlands. This has become part of the understanding of the relationship between mining communities, local cities and foreign cities (for example Bradbury, 1979 and 1980). The picture of a chain or hierarchy of linked localities and classes, exhibiting an unequal pattern of relations emerges, although explicit statements encompassing this whole thesis are rare.

The notion of the 'staples trap' is a central concept for some staple theorists. This is the suggestion that an economy which is primarily shaped to extract and export resources, develops and refines the institutional apparatus and skills which support these activities. This process distorts the economy and may inhibit, or even foreclose, any chance of the economy changing direction or moving to a new phase. This idea must be contrasted with the aim of many governments, including that of Western Australia, which seek to establish locally a stable, diversified, independent economy based on manufacturing with a significant role for local capital.(3) Regions which are located into a 'staples trap' effectively work in the interests of transnational companies since they are locked into a situation of economic dependence.

Support for the notion of metropolitan dominance comes also from a third body of literature, namely theories of imperialism (Brewer, 1980). Although mainly developed to explain economic development in the Third World, ideas generated in this literature have been transplanted directly or with some modification. As Edel and others (1978) point out, the least satisfactory of the transplants are those which interpret uneven regional development in capitalist economies in ways analagous to a process of 'internal colonialism'. By this logic, metropolitan dominance is explained as the result of oppression. Why and how the city, or agents located in the city, can oppress hinterland regions is not clear.

World system and dependency theorists have produced more influential analogies. Such theorists describe centre-periphery relations (or, in the case of Wallerstein (1974), the core periphery and semi-periphery relations) at an international scale in terms of the logic of exchange or circulation. A geographical specialisation based on an areal division of labour has evolved as a part of capitalist relations and international trade. The theories emerged from analyses of the role of the Third World in the international order, but the definition of centre and periphery has not been limited to the international level or to the relationship between the industrialised metropoli of Europe and North America and the Third World. Frank (1969) is well-known for his notion of a chain of metropolis-satellite relations in which relations of dominance and surplus extraction exist not only between direct producers and the immediate exploiters but at all levels in the world system.

> Indeed, it is this exploitative relation which in chain like fashion extends the capitalist link between the capitalist world and national metropolises to the regional centres (part of whose surplus they appropriate) and from these to local centres and so on to large landholders or merchants who expropriate surplus from small peasants or tenants, and sometimes even from these latter to landless labourers exploited by them in turn. (Frank, 1969:7-8).

Five major distinctions between centre and periphery can be distilled from the descriptions of the causes and consequences of dependency and world system relations. Many of these are reminiscent of ideas expressed in staples theory. Firstly, the pattern of unequal exchange is to the advantage of the metropolis which is in a position to extract surplus. Secondly, the centre has an established industrial base and an advantage in capital, skills and technology. (For some theorists, this implies autarchy, for others it is simply the conditions for a superior trade relationship.) Thirdly, centre and periphery can be distinguished in class terms, with the owners of capital located in the centre, exploiting peripheral workers. In other words, class and geography are assumed to coincide. Fourthly, the actions of the state operate to the advantage of the centre and in fact, the political realm, like class relations, is shaped to accommodate the patterns of unequal regional relations. Finally, the metropolis and metropolitan based classes enjoy higher wages, higher investment levels, and so on.

The combined weight of the various theses built around centre-periphery notion, (neo-classical, staples and imperialist) supports the temptation to interpret development in Western Australia in terms of centre-periphery relations both within the State and at national and international levels of analysis. Such a model seems at first to provide a simple and elegant expression of the patterns present.

Doubts about the value of using a centre-periphery approach for understanding urban and regional relationships in Western Australia are raised by critics of the models. Five points are worth noting here. Firstly, it is not clear that dependency theories can be successfully applied to understand the tendencies affecting advanced capitalist economies although this problem has not stopped Frank (1975) or a number of Canadian writers who have grafted them onto staples theories, from doing so. Second, a logic which is appropriate for analysing development at an international scale may be quite misleading at the subnational or regional scale. This reservation has been argued by Edel et al. (1978) who point out that regional peripheries within centre nations involve conditions such as low wage rates which contradict the assumptions used, for example by Amin (1976), to explain the existence of peripheral nations. Again, however, such problems have not prevented both Amin and Frank from arguing that dependency and centre-periphery relations can be used to explain regional disparities. (See Edel et al. 1978:6; Brewer, 1980:176-197).

Thirdly, many of the theorists treat the state as promoting metropolitan bias. The state however can operate in other ways as demonstrated in the now widely cited study of Alberta and Saskatchewan by Pratt and Richards (1979). The regional parochialism and province-building activities of these two active, interventionist, entrepreneurial governments in Canada have attempted to counter the metropolitan dominance of Central Canada, leading to the suggestion (as yet poorly supported) that there has been a power shift west. The role of the state in shaping the meanings of regional spaces and thereby affecting the activities of the private sector has been addressed theoretically in a recent article by Lipietz (1980). His most provocative suggestions are that the state acts both to impose a 'capitalist logic' on the legal framework defining land ownership and use, and to make up for the lack of a 'law of value in space' via planning, intrastructure investment and the like. Fourth, the dependency theories in general have been criticised for obscuring class relations by making the centre synonomous with the location of the ruling class and the periphery the locus of workers. The local bourgeoisie is accommodated by reference to a 'comprador class'. The definitions of particular classes and their relations in the context of regional geographies are not well understood. Hence, the assumption that class and geography coincide seems to be a crude, if convenient, simplification.

Finally, much the most critical point and one which is used to explain conditions in relation to Perth later in this paper, is made by Brewer (1980). Brewer points out that capital is not homogenous. The logic which governs the activities of local merchant capital may find its spatial expression in metropolitan dominance as a result of activities associated with collection, storage and redistribution of commodities. This is the basic premise of central place theory. The same logic need not apply to all fractions of capital. It is, in fact, quite unlikely to apply at the regional level, to the activities of monopoly capital which operates transnationally. Brewer's own words on this point are graphic.

> Merchant capital tends to create a geographical hierarchy where it is gathering together the output of scattered production units for export in bulk, or for bulk supplies to urban centres... Modern monopoly capital is often administered through a superficially similar geographical hierarchy. The head office is in a major world centre, regional offices or local subsidiaries are established in large cities, while productive activities are located wherever manpower, markets and raw material supplies dictate. This hierarchy of administration is not, however, in any real sense a 'chain of metropolis-satellite relations'. The intermediate levels of management are no more than agents of the corporation; they have no independent economic base. The real relation of exploitation is a direct wage relation between the workers and the corporation as a unit of capital (Brewer, 1980:173).

This obviously suggests that our research should be sensitive to the effects of disaggregating the analysis - not just for different fractions of capital, but also for phases in accumulation or the articulation of different modes of production. Once this is done it allows for new possibilities. Castells (1978) discusses the existence of forms of dependent urbanisation and identifies three types based on historical phases: (a) colonial dominance, (b) capitalist-commercial domination and (c) imperialist industrial and financial domination. The 1978 description is primarily based on Third World illustrations but Harvey (1973) argues that the same phenomenon occurs in any country or region where:

> the urban form exists as a channel for the extraction of surplus from a rural and resource hinterland for purposes of shipment to the major metropolitan centres. (Harvey, 1973:232).

He suggests such forms were characteristic of early nineteenth century American cities.

The focus here is not on the activities of productive, industrial or mining capital but on merchant, finance or commodity forms. As such, we are directed to look at the organisation of all aspects of the accumulation process in terms of their different spatial implications and in this some disaggregation of the fractions of capital will obviously produce a richer picture of the pattern of dependence relations. Work on the role of merchant capital in maintaining structures of underdevelopment in the Third World is evidence of this (Slater, 1978:32).

Equivalent work on uneven regional development in the western world is only now beginning to come out in English. European theorists such as Lipietz (1980) argues that the organisation of space is a product of (a) the interface between fractions of capital; (b) the current phase of capitalism and earlier phases of pre-capitalist modes; and (c) state intervention which shapes the meaning of space.

THE ROLE OF THE STATE IN RESOURCE DEVELOPMENT IN WESTERN AUSTRALIA; INFRASTRUCTURE POLICY

The contemporary boom in mining in Western Australia dates from the lifting of the federal embargo on iron ore in 1960. The decade of the 1960's was an iron ore and nickel boom focussed on the goldfields and Pilbara. In the 1970's the focus shifted to gas from the North West Shelf, bauxite in the south-west and potential diamond production in the Kimberleys. With the exception of a brief interregnum (1972-74) a Liberal-National Country Party coalition has ruled throughout the period with Sir Charles Court playing the major role in defining the government's development strategies, firstly as the responsible Minister and later as Premier.

There are two keystones to the Liberal policy, both of which have had important repercussions on Perth. The main instrument the government has used to encourage mining and industrial investment in the State has been an incentives package combining (a) limited royalties, (b) State control of labour and community opposition, and (c) State subsidies to production - notably in the form of energy pricing and provision of infrastructure. This section will focus specifically on the implications for Perth of the State's infrastructure policy. The other keystone to the policy has been the notion of multipliers, linkages or spin-offs from resource development. Western Australians have been persuaded that while the main mining projects may not produce many jobs and may initially be costly - ultimately there will be a payoff in the form of a highly expanded job market, higher personal incomes and a strong manufacturing sector. These two keystones, the package of investment incentives and the ideological use of the multiplier, are the State's response to the competing and often contradictory pressures of accumulation and legitimation, or the need to simultaneously accommodate both the interests of companies and of voters, (Harman, E. 1981b).

The State has historically provided basic social infrastructure. Schemes which can be directly tied to the needs of a particular mining company, however, probably have their origins at Wittenoom in the 1940s when the State under the initiative of a senior bureaucrat (Dumas) financed the construction of 150 houses for Australian Blue Asbestos (Layman, 1981). Not all projects have had this type of support. The guiding principle used by Court has been one of pragmatics. If the companies will invest while still picking up the tab, well and good. The iron ore projects of the 1960s in the Pilbara obstensibly involved limited State outlays, being based on the terms of the development agreements largely on company-financed towns, rail and port facilities. Yet in the 1970s the State again proved more willing to finance infrastructure, this time to support the North West Shelf agreement. (The factors underlying these shifts are discussed at the end of this section). In practice, projects which are more remote or which are judged to be marginal are given State support.

The infrastructure policy is costly and two tendencies derived from these costs have become increasingly evident in Perth.

a. There has been a shift in the proportion of public funding for basic services out of the metropolitan area to the advantage of the resource regions; and
b. the price of major domestic services like gas, electricity and transport has risen sharply coincident with the government taking on large resource-related infrastructure commitments.

The Australian Institute of Urban Studies (1980) was the first to suggest publicly that the metropolitan areas were being squeezed by resource development. They have since been echoed by the Real Estate Institute - an organisation which obviously has an interest in protecting the level of amenities and services provided by the State especially in the cities where major real estate investment occurs. Neither body supported their assertions with detailed figures on the expenditures involved. Figures 1, 2 and 3 provide evidence for Perth.

It should be noted at the outset that accurate information on the changing distribution of public expenditure in the State (by sector and region) is not easily to come by in Western Australia. Most of the data used here are taken from Australian Bureau of Statistics figures and Annual Reports of relevant departments. Further work is being pursued using budget papers and other sources. The ambiguities surrounding the infrastructure programme and related expenditures have been created in part by the structure of the Western Australian bureaucracy. The co-ordination of policy relating to major projects is handled by a non-statutory body, the Planning and Co-ordinating Authority chaired by the Department of Resources Development. Financial aspects are handled directly by Treasury. A sub-committee of the Planning and Coordinating Authority, the Townsites Development Committee oversees (at least nominally) the greater part of State expenditure or social infrastructure in the towns. The activities of this group lie completely outside public scrutiny (see Bowman, 1979, Chap. 4; Government of Western Australia, Department of Town Planning, 1982).

FIGURE 1 Metropolitan Share of Public and Private Building, 1968-1980.

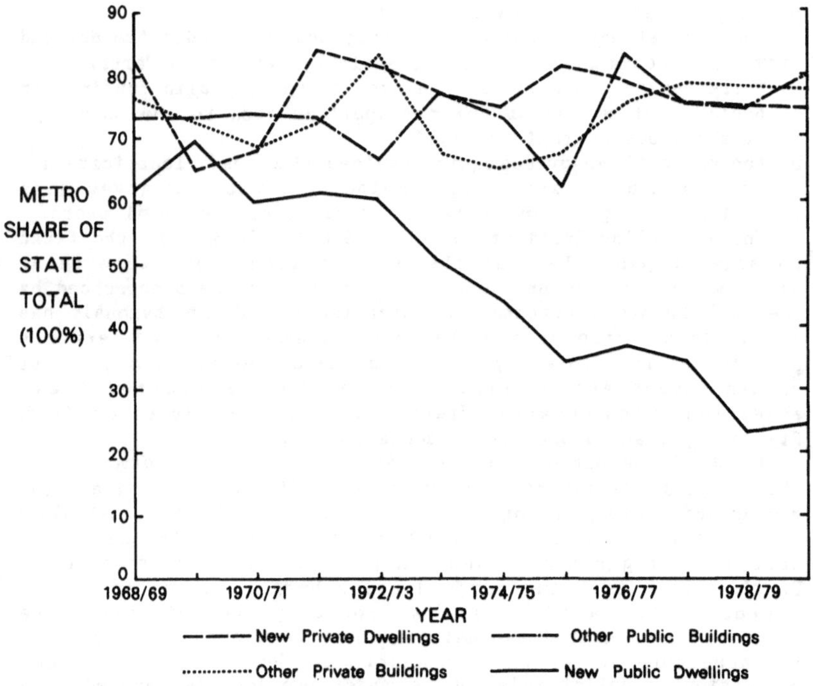

FIGURE 2 Metropolitan Share of Housing Units constructed by the Western Australian State Housing Commission, 1944/45–1979/80.

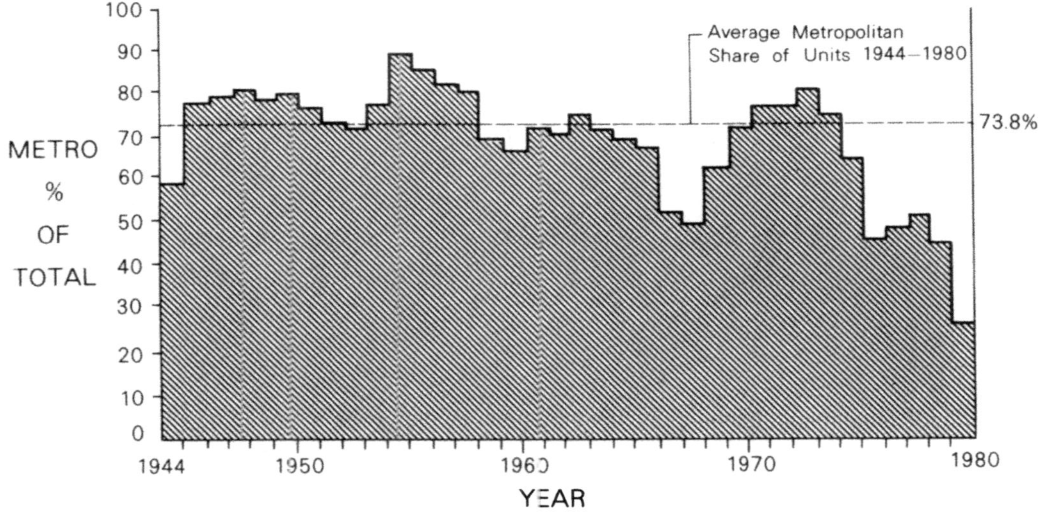

Source: State Housing Commission, Annual Reports.

FIGURE 3 Metropolitan Share of Infrastructure Expenditure by Department, 1959/60-1980/81.

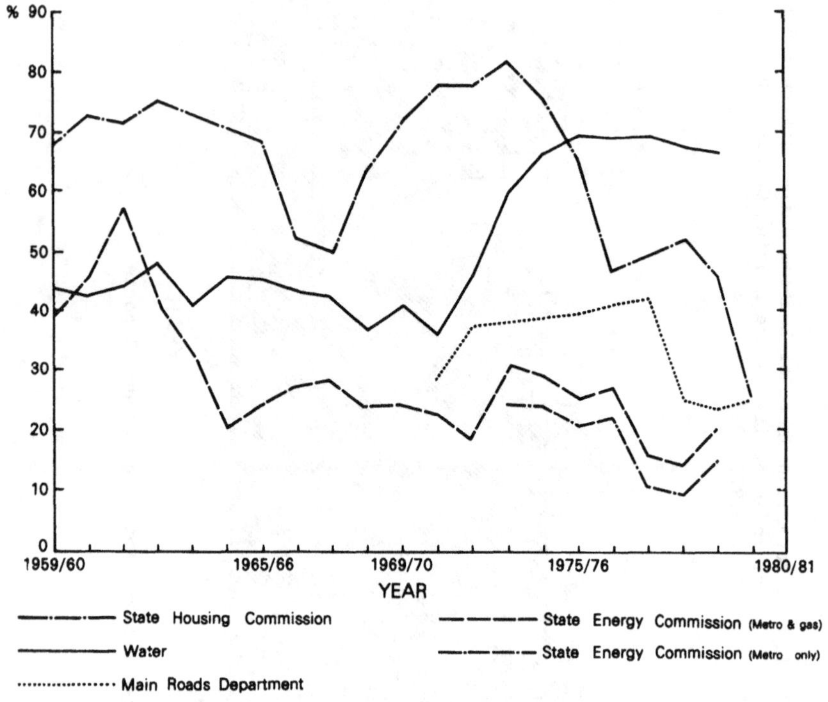

Source: Departments, Annual Reports.

Figure 1 is based on building approvals and refers specifically to building programmes. It illustrates the proportion of state-wide expenditures made by the public and private sectors in the metropolitan area between 1968/9 and 1979/80. In both cases, dwellings are graphed separately from other types of buildings. It is clear that in the last decade public housing in the metropolitan area has received a declining share of public funds, capturing almost 70 per cent in 1969-1970, but less than 25 per cent in 1979-80. This is clearly the outcome of deliberate policy decisions and not simply an aberration which has affected all construction. It might, for example, be argued that resource projects have bouyed up non-metropolitan economies and diverted the market as a whole. However, this is clearly not the case. Throughout the decade, the private sector has committed a constant 65-85 per cent of expenditures to Perth.

The State Housing Commission is the major public agency involved in housing construction and provides annual figures on its activities. As Figure 2 illustrates, in the thirty-five years which have elasped since its inception, the State Housing Commission has built, on average, almost three quarters of its new units in the Perth metropolitan area. There have been two periods when the city's share has fallen significantly. In the second half of the 1960's, coincident with the nickel and iron ore booms, the percentage in the metropolitan area dropped to around 50 per cent. And again, since the mid 1970's coincident with projects related to bauxite and gas, Perth's share has fallen to the point where in 1979-80, it received just over a quarter of new State Housing Commission units. This contrasts strongly with the peak in the first half of the 1970's when the Whitlam government's commitment to urban development, and the rapid growth of the city itself, both had a positive effect on the metropolitan housing programme.

It is easy to offer explanations other than support for resource projects to explain the fall in the metropolitan share of State Housing Commission funds during the second half of the 1960's and 70's. The Commission took on responsibility for Aboriginal housing in the mid 60's which diverted funds into rural areas. Both the Fraser and Court Governments have visibly reduced the priority given to social and urban welfare in the late 1970's. In addition, the State Housing Commission has suggested that a large number of flat vacancies in public housing is evidence of an oversupply in the metropolitan area and hence justifies a shift of funds elsewhere. While they have certainly contributed to the decline in metropolitan funding, these alternative explanations do not account for the magnitude of the metropolitan loss. Funds have been specifically earmarked for supporting resource developments. For example, in 1970 a major recipient of State Housing Commission funding in direct support of a major project, was Port Hedland. Eighty-six houses were built by the government and a new town (South Hedland) created to house workers associated with the port operations of Mt. Newman and Goldsworthy. On 1st April 1978, $20 million was specially earmarked for a northwest housing programme to provide facilities in the Pilbara and the Kimberleys. This was described as part of the State's planning to cope with demands associated with major resource projects. Other resource based communities in which the State has provided housing include Karratha, Wickham, Koolyanobbing and Capel. Since the State Housing Commission's role has been more active in 'open' towns, state housing is likely to play a greater rather than a lesser role in the future. The government is no longer enthusiastic about the company or 'closed' towns typical of the 1960s.

Figure 3 graphs the trends in the metropolis' share of expenditure in other service areas. The city's share of expenditure on water and sewerage has shown a marked improvement over the period - a fact which may be accounted for both by the urban priorities of the Whitlam government and by the increasing pressure on Perth's water supplies in the 1970's. A slight but discernable decline in the share of road, electricity and gas expenditure is apparent. It is useful to keep in mind that Figure 3 is a conservative estimate of the shift in expenditures out of Perth. The infrastructure programme was given it's greatest impetus in the second half of the 1970's and major expenditures are not yet evident in Annual Reports. It is clear that the major resource-related commitments made, for example, by the State Energy Commission do not appear here. The Dampier-Wagerup gas pipeline, Pilbara power integration scheme, the Kwinana power station, conversion and extension of a station at Muja, all involve largely future capital expeditures on a massive scale for which the government has approval from the Federal Loans Council to borrow heavily on overseas money markets. The expenditures involved are given in Table 1.

TABLE 1 Western Australian Loan Council Borrowings, 1980-82

Purpose	Total Sought	$m Approved for 1980/81	$m Approved for 1981/82
Dampier-Perth Pipeline	553.3	9.2	35.5
Pilbara Power Integration	138.7	3.7	5.9
Worsley Rail & Water	46.2	7.1	1.6
N.W. Shelf Gas Infrastructure	38.6	13.2	4.8
Kwinana Power Station Conversion	39.1	5.1	19.0
N.W. Shelf Gas Infrastructure - Jervoise Bay	7.05	6.3	1.6
Muja "D" Power Station	236.3	12.2	52.8
Totals	1055.6	56.8	121.2

Source: Western Australian Government Treasury.

The government's commitment to major infrastructure expenses has coincided with cutbacks in social services generally (education and health) and sharp increases in the prices charged to consumers for many public services during 1981. The Labor opposition made considerable political mileage out of increases in State government charges which they calculated as rising much faster than the rate of inflation. From 1980/1981 to 1981/1982 the average family faced an average weekly bill increase for state charges of $10, from $31.11 to $41.1 (M. Naylor, personal communication). Rising costs and cutbacks in services like education funding were interpreted as a product, in part, of the carrying costs associated with the infrastructure policy and un willingness by the Liberals to levy significant mining royalties. However, the government itself charged that the rises must be laid at the feet of the Federal government's constraints on finance and notably the actions of the Commonwealth Review of Functions (commonly known as the 'Razor Gang'). The intricacies of state budgeting and federal-state financing make it difficult to apportion the blame in any precise way. Nonetheless, it seems clear that the public costs associated with the infrastructure policy are hitting consumers, notably metropolitan consumers, very hard at present. The issue surfaced again in 1982 with fears that the postponement of the second phase of the North West Shelf will mean that local consumers will pay a high price for the gas for which the State Energy Commission is committed.

This section has provided evidence (admittedly partial and limited) that seems to suggest there has been a shift in the distribution of public expenditures on infrastructure over the period of resource development. Prior to 1960 there was already a predilection to use State monies to provide supporting services and infrastructure for major development initiatives - the State's investment in Wittenoom in the 1940's and Kwinana in the early 1950's are cases in point, though the history of such intervention is earlier than this (Bolton, 1981). This role was stepped up with the onset of the contemporary mineral projects. Brown (1981) has argued that increased public expenditures on resource areas are evident even in the 1960s when the major development agreements seemed to leave most infrastructure financing in the hands of the private sector. Certainly by the 1970s the Liberal government was quite explicit in its pursuit of funding to support an active role in infrastructure provision. This shift is recognised in departmental reports (for example, the report on resources and town planning - Government of Western Australia, Department of Town Planning, 1982:17).

As a result of this, there appears to have been a shift in the pattern of public expenditures away from services and facilities required to meet welfare needs in favour of those used to support the needs of labour and capital in the major mining areas. To use O'Connor's categories (O'Connor, 1973) this has meant a shift out of social consumption (policies contributing mainly to the social and material support of the population) towards social investment (policies contributing mainly to private sector profitably through provision of the necessary means of production).

There are at least three sets of factors which any explanation of this shift in the infrastructure programme must recognise and each level of explanation is embedded in the next. First, there is need to recognise the frontier nature of Western Australia's geography and history of development. It was until 1950 largely agrarian and until the 1960s was dependent on federal revenue-sharing arrangements. Industrialisation at Kwinana from 1952 and the mineral boom since, have wrought a dramatic change in thinking in the State (Layman, 1982; Harman, E., 1981b). Mining is seen as the key to the State's industrial development, its economic viability, levels of public and private wealth, and to the opening up of vast areas of relatively uninhabited territory. The package of investment incentives including the infrastructure policy, is a concrete expression of the partnership now forged between the State government and large-scale mining interests in Western Australia (Harman, E., 1981b). Galligan (1981) has called this the 'rentier' phase in resource policy and contrasts it with an 'entrpreneurial' phase in which governments are much more committed to the collection of resource revenues and the use of such revenues for public investment in productive activities.

The second point to note is that the mining developments in the State since the 1960s have been mega-projects involving major Australian and transnational companies. The capital controlled by such companies is highly mobile and the companies have become remarkably adept over the last quarter century in creating the notion that there is an international capital shortage and forcing governments to compete with each other for investment. The infrastructure policy has evolved in Western Australia (as it has elsewhere in Australia and Canada) as a means of underbidding other jurisdictions.

The third and final point is that the priority given to the infrastructure policy has been greatly strengthened in the 1970s in the context of the international economic crisis and the belief that profit rates are falling. By the mid 1970s the Western Australian government felt obliged to contribute to infrastructure costs to offset a decline in profit rates which threatened in the words of Court to "prove Marx right" (Court, 1978) and which brought the average return in the mining sector to a "dangerously low" 12.6 per cent by 1980 (Court, 1980). The increasing tendency for the state to provide basic infrastructure in advanced capitalist societies has been widely discussed in literature on the role of the state in general and in cities and settlements in particular, (see for example, O'Connor, 1973 or Saunders, 1981, Ch. 8). This practice is seen as both a cause and a consequence of the crises of capitalism. Hence, what was originally conceived in Western Australia by Dumas and Court as a pragmatic policy designed to cope with conditions associated with specific projects, can now be seen in terms of international trends affecting advanced capitalist states.

The shift in public expenditures from social consumption

(welfare) to social investment (support of production) has an important geographical corollary. Since most of the supporting services are required in resource-rich regions, there has been a diversion of funds out of the metropolitan area relative to the mineral-rich localities. There are, of course, significant exceptions. Leave was sought from the Loan Council in 1980 to borrow some $7 million, in addition to departmental funding, to invest in the development of a suitable industrial site at Woodman's Point in the Cockburn Sound industrial area for the manufacture of modules for the North West Shelf project. For the most part, however, public funds for resource projects have been spent outside Perth.

This would appear to contradict the logic of dependency models which view the state as promoting metropolitan dominance. The diversion of public funds out of social consumption and from Perth is evidence of the extent to which the position of most urban-based classes (welfare, working and middle-class groups) has weakened over time relative to capital-owning classes with interests in major mining ventures and auxiliary industries. That the majority of these are not located in Perth is evident from knowledge of the transnational ownership structures of most of the major companies and from evidence of income leakage out of the state (Harman, F., 1981).

There is one conceivable way in which the diversion of public funds, loss of services and increased costs in Perth could still be seen as ultimately and indirectly contributing to metropolitan dominance. This is the situation where the outflow of public funds and rise in costs is more than compensated for in the form of return flows of economic activity in the secondary and tertiary sectors via the multiplier and spin off. This is the prime benefit which Western Australians have been led to expect will flow to them from mineral development. To promote dominance, there must not only be a faster growth in the metropolitan economy vis a vis the regional economy - the growth must occur in key areas which will facilitate the extraction of mineral surplus by Perth-based interests (for example, the provision of expensive services to mining) or which promote the city's position as the centre of industry and innovation (for example, technologically advanced local industries). Whether this compensation is, in fact, apparent in Perth is the question addressed in the next sectiion.

PERTH AS A COMMERCIAL CLEARING HOUSE

According to the Liberal government the benefits of mineral development will accrue to the present generation (over 70 per cent of whom live in Perth) in the form of new job opportunities and higher incomes generated by multiplier effects. Benefits to later generations will exist in the form of a stable, diversified and relatively autonomous economy, including a strong manufacturing sector. Implicit in the arguments made by Sir Charles Court is an underlying theory of development which envisages the economy moving through three stages: (a) an export-orientated staples economy dependent on foreign capital and skills. This shifts towards (b) a diversified second stage when multipliers, linkages and externalities encourage the growth of other sectors, and which transform the extractive base towards resource upgrading and manufacturing, and (c) the final stage of economic independence or autarchy when the local economy has acquired its own indigenous capital and skills (Harman, E., 1981b, Figure 4:185).

Government members argue that their strategy is having considerable success. They point to estimates which have put the local employment multiplier for bauxite-alumina and for iron ore at 4.9 and the household income multiplier at 5.5, (Government of W.A., Department of Resources Development, 1980). They point also to the proportion of contracts and purchase orders which the main mineral developers have let in the local economy. The estimates range from 55 per cent of Worsley orders; 57 per cent of Woodside; 63 per cent for the Hamersley Iron concentrator; and 86 per cent of Wagerup orders (Harman, E. 1981a:338).

The health of the Western Australian manufacturing sector is a key to the whole discussion - not only because it is the pivot of the government development policy and an important means by which benefits should be returned to Perth, but because it has been seen as an important part of the definition of a dominant centre in most uses of the centre-periphery notion. For the world system, dependency and staples theorists, for example, the centre (or core) areas are the industrial heartlands and enjoy superior levels of technological development, entrepreneurship and economic independence.

Manufacturing in Western Australia has been, and still is, overwhelmingly concentrated in Perth (84 per cent of employment), but the sector has always been very small employing only 12.5 per cent of the labour force in the state compared to 19.7 per cent in Australia as a whole (1980 figures). It has largely focussed on the processing of local resources (agricultural, forest and mineral products). Far from seeing the growth of a strong, diversified sector in Perth over the last two decades, manufacturing has : (a) declined in importance; (b) become even more specialised and dependent on the export sector and (c) the regions, not Perth, have received the larger share of growth in the metals related sectors.

The decline in the relative importance of the manufacturing sector is evident in employment figures. The 12.5 per cent of the labour work working in manufacturing in 1980 represents a decline from 18.8 per cent in 1966. The decline is, of course, relative. There has not been an <u>absolute</u> loss in jobs as there has been in the eastern states, but in fact, a small increase. But this comparison is misleading in that it ignores the very severe international pressures which have led to the eastern sector being restructured and firms moving off-shore to cheap labour and tax havens in Asia. Moreover, the increase in manufacturing jobs in Western Australia has been tiny; a mere

4000 jobs providing less than 2 per cent of the more than 200,000 new jobs created between 1966 and 1980.

Not all manufacturing groups have performed identically. A specialised metals-related sector has emerged encouraged by mining activity. Employment in establishments providing basic metal products shows the most significant increase, moving upwards by almost 50 per cent over the 1970s. Growth in the metals area comes from two distinct sources. The major mining companies have invested in plants to undertake the first step in resource upgrading (metallising of iron ore, refining of bauxite to alumina, and the refining of nickel). At the same time, some local engineering and fabricating shops in the State have picked up additional work to service the major projects.

The emergence of a specialised metals related group is not, however, necessarily a step towards diversification and economic dependence. The group is more, not less, dependent on the actions of the major transnationals controlling mining. The processing plants are owned by the same companies and are a part of their vertical integrated structure. In addition, the metal-fabricating shops are very small in size. The average number of employees in establishments handling non-metallic mineral products is 43; in basic metal products 133; in fabricated metal only 24, in transport, 42; and in other machinery 33. As Schaffer (1981) notes with respect to a very similar situation in Alberta, these establishments are also highly dependent on the economic health of the export sector and the procurement policies of the transnationals.

Nor have the forces shaping the manufacturing sector encouraged metropolitan dominance. Perth has not been the major beneficiary of that growth in manufacturing which has been stimulated by the mining sector. In the first place, most of the resource up-grading is done in the regions and Perth's share of employment in most of the metals-related classes has fallen slightly over the last decade. Moreover, while manufacturing accounted for 23 per cent of metropolitan employment at the 1961 census, by 1976 it had fallen to just over 16 per cent (although it still undoubtedly dominates as the main location in the State).

The South-West Statistical Division now has close to 10 per cent of the State's manufacturing employment and important mineral related activities such as minerals sands processing and alumina production in the vicinity have helped make Bunbury a growth centre now surpassing Kalgoorlie as the State's second city (Government of Western Australia, Dept. of Industrial Development and Commerce, undated). Even accounting for the very different employment bases, the growth of employment in metals-related products in non-metropolitan Statistical Divisions like the South-West, Midlands and Pilbara was spectacular between 1966 and 1976 compared to Perth - annual rates of 7.1, 22.5 and 17.5 compared to 0.02 per cent respectively (McKeon, 1982:80).

An analysis of the procurement policies and contracts let by major developers like Woodside provides some additional clues as to why Perth may not be capturing as many of the spin-offs to manufacturing as is popularly supposed. (For a detailed discussion see Harman, E., 1981a). In the first place, estimates which give a precise figure to the amount of goods and services which are locally sourced (e.g. 57 per cent for Woodside) are highly suspect. They are guesstimates - albeit educated ones - rather than valid and reliable quantitative measures. The term "Western Australian Contract" implies that a job has been given to a local firm and all the money will be spent locally, using local labour. In fact, firms which have been given contracts called West Australian in total or in part, include genuine local enterprises, long established branch offices and subsidiaries of eastern states or overseas firms; newly established branches or subsidiaries; and contract offices handling the job on a turn-key basis. Joint ventures and consortia complicate the picture still further.

Of 36 major contracts over $1 million let by Woodside up to late 1981, 27 were designated by the company as local in total or in part. Of these, some 23 had out-of-state parents (see Harman, 1981a, Table 5:343). It is impossible to know without knowledge of the internal operations of such companies just how much autonomy in decision-making such firms have or to what extent they use and develop local skills or reinvest locally. It is equally difficult to ascertain whether the several joint ventures involving a local partner with a transnational will lead to any technology transfer as is assumed by Court's developmentalism and as would be required if Western Australia or Perth is to develop an indigenous base for technical innovation. The common practice is in fact for the local firm to provide labour, goods, services and for the non-local partner to provide specialised skills, technology and the like with no necessary transfer of expertise. A case in point is the ETPM-Clough joint venture handling off-shore pipe-laying where Clough Engineering plays the local role and ETPM is a French firm contributing technical know-how, specialised equipment and the ship.

From the point of view of manufacturing, only 5 of the 27 jobs involve fabricating - most are construction or service jobs, although these include some engineering and design work. From Perth's perspective, most of the activity is concentrated outside the metropolitan area, notably in the Burrip Peninsular, although Karratha and Geraldton also feature. Some of the contradictins in this situation have been evident in the complaints by the Confederation of Western Australian Industry (representing local companies) that insufficient contracts have been let in the city. This has put local capitalists in alliance with the Trades and Labour Council also making representations to the government to get more local work from project contracts.(4)

FIGURE 4 Employment change by industry sector, Western Australia, Civilian Employee Series, 1971-1980 by percentage.

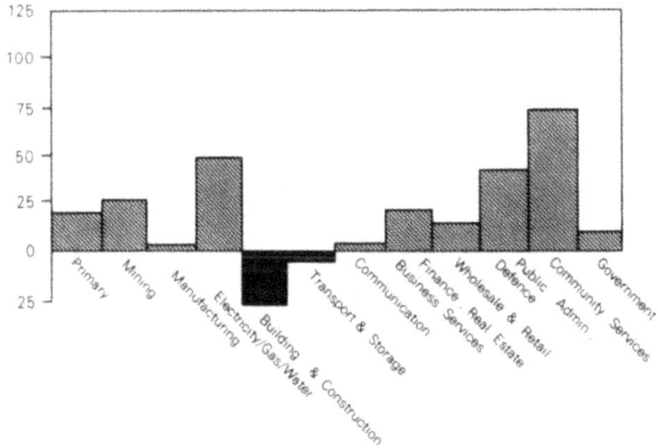

Source: Australian Bureau of Statistics.

FIGURE 5 Employment change by industry sector, Western Australia, Census, 1961-1976 by percentage.

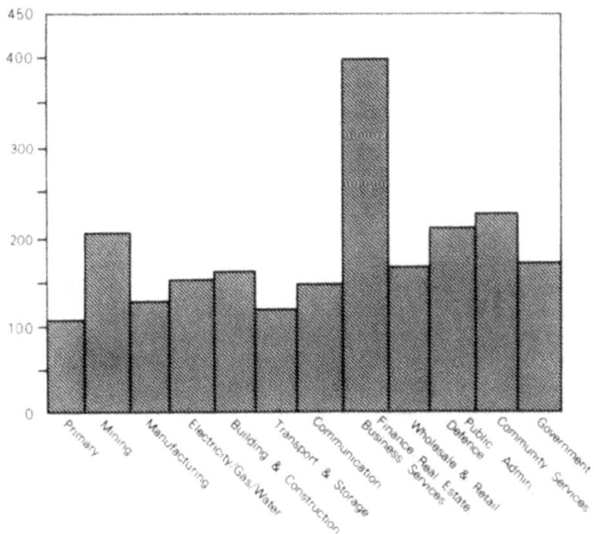

Source: Census.

While the picture for manufacturing may be a little bleak, the same is not true for the tertiary sector in Perth. Over 80 per cent of new job creation in the State since 1966 has been in the tertiary sector with the private sector providing some 98,000 new positions (of a total of more than 200,000) while the public sector contributed a further 73,000. In fact, the rate of growth was fastest in public sector jobs and is evident in the strong increase in employment in public administration and defence, community services, electricity, gas and water (Figure 4). The rate of growth in privately supplied services differs according to source and time period. But it would appear that finance, real estate and business services have shown the strongest growth since the 1960s, outpacing even the growth in public sector classes according to census changes, (Figure 5). The detailed breakdown of tertiary employment is not available for all classes in Perth unfortunately. However, over 70 per cent of tertiary employment exists in the metropolitan area and the expanding skyline has mirrored the unprecedented expansion of office space notably from the late 1960s and again in the early 1980s. The popular conception that this expansion is directly attributable to mining is reinforced by the prominence of Mt. Newman House, Hamersley House and Allendale Square (Woodside) on St. George's Terrace.

To what extent the growth in services can be explained as an outcome of mineral development is difficult to estimate precisely. The government's own input-output analysis of the alumina-bauxite industry concluded that the multiplier effect was much the greatest in services (Government of Western Australia, Department of Resources Development, 1980). It is also now recognised that an export-led service sector can provide the foundation for trade in newly developed skills and services. The problem is that the tertiary sector has been taking an increasing proportion of Australian jobs throughout this century, with the real impetus coming from the expansion of quaternary or information-processing activities (Wilson, 1980:278). The only evidence we have that contemporary mineral projects may now be giving this national trend an extra push in the resource-rich States, is by looking at comparative figures.

McKeon (1982:27) used census figures (1966, 1971 and 1976) to demonstrate that while all other States declined in their share of Australian employment in the tertiary sector (or held even), Queensland increased its share from 14.1 per cent to 15.1, while Western Australia went up from 7.5 to 8.8 per cent. Likewise, Jarvie (1981) used the census to show changes in tertiary employment by metropolitan area and also showed that while the percentage change in New South Wales and Victorian cities between 1971 and 1976, was 7.0 and 12.13 per cent respectively, it was a much stronger 19.8 and 20.9 per cent in Queensland and Western Australia. The problem with all these figures is that the increased share of tertiary employment taken by the peripheral resource-rich states parallels a similar increase in population shares. Both, of course, may be explained by expectations and needs associated with mining developments.

While the evidence is clearly circumstantial, it does suggest that Perth may by handling an increasing level of administrative services for the government, private sector and civilian population, directly and indirectly related to mineral development. For the major mining companies, the city is a commercial syphon or clearing house for the paperwork surrounding the inflow of investment destined for new projects and the outflow of commodities, purchase orders or profits. Major decisions may still be taken out of State, orders ultimately sourced and profits relocated externally. But all of this still requires brokerage services in the local metropolis and hence the expansion of office space and employment. Property services are also likely to have increased to meet both the real and speculative demand for residential and commercial real estate. At least part of the growth in the public sector may be attributed to the greatly expanded role of departments and agencies involved in supporting the major projects (for example Treasury, Department of Resources Development, State Energy Commission, State Housing Commission). To ascertain the veracity of any of these speculations would require much more detailed historical analyses of local banking and other service activities and/or econometric analyses. Nonetheless the suggestions paint a reasonable scenario.

CONCLUSIONS

If Perth does play the role of broker or commercial service centre, then it is functionally tied to mineral developments in the regions. And while city and region each have a specialised role to play, it is difficult to establish substantive grounds for calling Perth dominant or especially advantaged in this arrangement. Instead, the city qualifies for the description of some form of dependent urbanism. Perth plays a necessary role in circulation relating to mining in the regions, but a limited role in production. Perth-based firms, offices and agencies handle the movement of both money transactions and shipment of other commodities (that is the inflow of private and public investment and the outflow of profits, and the inflow of purchase orders and outflow of minerals and mineral products). Firms engaged in other activities related to the resource sector are limited to some fabricating, engineering, design work and construction. In their role as siphons, city-based interests perform a relatively low level function and have no real power to extract surplus. Union workers in the resource regions have more power to extract a share in the form of higher wage demands than city service workers, in both the private and public sector many of whom are women in low-paid non-managerial positions. (The power of Western Australian mining unions is still not sufficient however to prevent wage levels being lower than eastern-state equivalents).

TABLE 2 Metropolitan and Non-Metropolitan Incomes by State

	Annual Personal Incomes 1976(1)			Mean Income, 1973-74(2)		
	Metro $	Non-Metro $	Metro as % of Non-Metro	Capital City $	Rest of State $	Capital as % of Rest of State
New South Wales	4977	3567	139.5	4220	3410	123.7
Victoria	5144	3399	151.3	4230	3730	113.4
Queensland	4496	3674	122.4	3670	3780	97.1
South Australia	4562	3777	120.8	3700	3780	97.9
Tasmania	4395	3687	119.2	3730	3650	102.2
A.C.T.	6475	4631	139.8	N.A.		-
N.T.	6750	4656	145.0	N.A.		-
Western Australia	4534	4562	99.4	3800	4250	89.4

Source 1. Census, 1976
2. ABS Catalogue No. 6504, 1973-74, (Based on survey of consumer income and expenditure).

This interpretation of Perth's role helps to explain why so many Perth residents have difficulty in matching their personal reality with the statements that Western Australia is booming, is rich or is the "State of Excitement". It does explain why, despite the fact that economic activity is generating wealth in the State, residents are receiving relatively low incomes per capita compared to other Australians (Harman, F., 1981). It explains why in this State (unlike others in the country) per capita incomes in metropolis and region are virtually equivalent (Table 2). In most states, urban incomes are much higher (a basic prediction of most centre-periphery models). It helps also to explain why the local social welfare agencies feel hard-pressed as is evident in the startling increase in the demand for emergency relief services since 1975 and especially during 1981 (Western Australia Department of Community Welfare, 1981).

In 1982, there seems little doubt that Perth is not a boom-town. Media headlines like 'Perth on a Tightrope after Resources Bust' (Financial Review, March 24, 1982:16) are testimony to this. But even prior to this, in 1980/81 when it was apparent that a great deal of investment was occuring in the State and wealth was being generated, the evidence of any boom in Perth was conflicting and ambiguous. On the one hand, there was an explosion in planned office construction and well-publicised sales of expensive prestige housing. On the other hand, the housing market overall seemed depressed in Western Australia, compared with the relative boom in building in Queensland (Indicative Planning Council, 1981).

The contradictions in this description dissolve if we accept that population growth is as much a function of the expectation of boom conditions as of any actual increase in long-term job and income opportunities (a condition reminiscent of the notion of the boom as a self-fulfilling prophecy advanced by Copithorne, 1980). The expansion in office space reflects both speculative expectations and the real increase in the demand for commercial services associated with Perth's role as a clearing house. The housing market is depressed especially at the middle and lower end of the price ranges by high interest rates and the fact that Perth residents have no special income advantage which would reduce this impediment.

The tendencies shaping this relationship between Perth and the rest of Western Australia since 1960, and the onset of the contempory mining booms, do not appear to have given Perth much more than an expanded skyline and a larger population. It is difficult to find substantive evidence that the city has established the kind of metropolitan dominance and pattern of regional dependencies depicted in the centre-periphery notions of dependency theory. The lack of dominance appears to have come about as a result of both State policies which have directed public funds out of the metropolitan area, and the operations of mining capital which do not confirm (except in superficial ways) to the logic assumed in the centre-periphery models.

NOTES

1. An earlier version of this paper was presented to the Pacific Regional Science Association, Seventh Annual Conference, Surfer's Paradise, August 16-20, 1981.
Research for the paper was assisted by a grant from the Australian Research Grants Council.
2. See The Canadian Review of Sociology and Anthrpology. Special issue on dependency, underdevelopment and regionalism, 17:3, 1980.
3. The logic of the Court government in Western Australia along these lines is outlined : detail in Harman, E.J., (1981b).
4. See for example comments in 'Work-flow in W.A. is Monitored - Govt.', The West Australian, (Perth), June 26, 1981, p. 10, and 'W.A. Loses Grip on Bonanza', The Western Mail, (Perth), June 13, 1981, p. 14, and 'T.L.C. Seeks Aust. Gas-Jobs Drive', The West Australian (Perth), September 23, 1981, p. 38.

REFERENCES

Amin, S. (1976) Unequal Development: an essay on the social formations of peripheral capitalism, Translated by Pearce, B., Brighton, Harvester Press.

Australian Institute of Urban Studies (1980) 'Urban Strategies for Australia: Managing the 80's', AUIS publication No. 88, Canberra.

Bolton, G.C. (1972) A Fine Country to Starve In. Nedlands, University of Western Australia Press.

Bolton, G.C. (1981) From Cinderella to Charles Court: The Making of the State of Excitement', in Harman, E. and Head, B., eds, State, Capital and Resources in the North and West of Australia. Nedlands, University of Western Australia Press.

Bowman, M. (1979) Australian Approaches to Environmental Management. Hobart, University of Tasmania.

Bradbury, J.H. (1979) 'Towards an Alternative Theory of Resource Base Town Development', Economic Geography, 55, 2:147-66.

Bradbury, J.H. (1980)'Instant Resource Towns Policy in British Columbia, 1965-1972', Plan Canada, 20, 1:19-38.

Brewer, A. (1980) Marxist Theories of Imperialism: A Critical Survey. London, Routledge and Kegan Paul.

Brown, J. (1981) 'Infrastructure Policies in the Pilbara' in Harman, E. and Head, B., eds, State, Capital and Resources in the North and West of Australia. Nedlands, University of Western Australia Press.

Castells, M. (1977) The Urban Question: A Marxist Approach. London, Edward Arnold.

Clements, W. and Drache, D. (1978) A Practical Guide to Canadian Political Economy. Toronto, Lorimer.

Copithorne, L. (1980) Natural Resources and Regional Disparities. Economic Council of Canada, Ottawa.

Court, C.W. (1978) 'The Road Back to World Recovery', A speech presented to the Int. Union of Young Presidents' Org., Sydney.

Court, C.W. (1980) 'Fight for the Future', Address to the Australian Mining Industry Council, Canberra, 9th May.

Edel, C. and others (1978) 'Uneven Regional Development', An Introduction to this Issue', The Review of Radical Political Economics, Special Issue on Uneven Regional Development, 10, 3:1-12.

Fowke, V. (1957) The National Policy and the Wheat Economy. Toronto, University of Toronto Press.

Frank, A.G. (1969) Capitalism and Underdevelopment in Latin America. New York and London, Modern Reader Paperbacks.

Frank, A.G. (1975) 'Development and Underdevelopment in the New World: Smith and Marx vs. the Weberians', Theory and Society, 2, 4:431-466.

Galligan, B. (1981) 'Federalism and Resource Development', Paper presented at State, Capital and Labour Conference, Murdoch University, Western Australia.

Government of Western Australia, (1981) Department of Community Welfare, departmental paper 'Emergency Assistance.'

Government of Western Australia Department of Industrial Development and Commerce, (1981) 'Manufacturing Industry in Western Australia'.

Government of Western Australia, Department of Resources Development, (1980) 'Major Development Project Impact Study, Economic Impact of the Western Australian Alumina Industry', June.

Government of Western Australia, Department of Town Planning, (1982) 'Resource Development and Urban Planning in Western Australia' Planning Minister's Conference, Perth, W.A. 18th-19th February.

Harman, E.J. (1981a) 'Mining and the Manufacturing Sector in Western Australia', in Harman, E. and Head, B. eds, State, Capital and Resources in the North and West of Australia. Nedlands, University of Western Australia Press.

Harman, E.J. (1981b) 'Ideology and Mineral Development in Western Australia, 1960-1980', in Harman, E. and Head, B. eds, State, Capital and Resources in the North and West of Australia. Nedlands, University of Western Australia Press.

Harman, F.J. (1981) 'Resource Development and Personal Income Levels', in Harman, E. and Head, B. eds, State, Capital and Resources in the North and West of Australia. Nedlands, University of Western Australia Press.

Harvey, D. (1973) Social Justice and the City. London, Edward Arnold.

Holland, S. (1976) Capital Versus the Regions. New York, St. Martins.

Indicative Planning Council for the Housing Industry (1981) Report for 1981-1982 to 1983-1984, Canberra, AGPS.

Jarvie, W.J. (1981) 'Internal Migration and Structural Change: Some Preliminary Observations'. Paper presented to the Seventh Pacific Regional Science Conference, Queensland, August 16th-20th.

Layman, L. (1981) 'Changing Resource Development Policy in Western Australia, 1830's-1860's', in Harman, E. and Head, B. eds, State, Capital and Resources in the North and West of Australia. Nedlands, University of Western Australia Press.

Layman, L. (1982) 'Development Ideology in Western Australia 1933-1965' Historical Studies, October, (in press).

Lipietz, A. (1980) 'The Structuration of Space, the Problem of Hard and Spatial Policy' in Carney, J., Hudson, R. and Lewis, J. eds, Regions in Crisis. London, Groom Helm.

McKeon, R. (1982) 'Patterns of Regional Development and an Examination of Possible Indicators of Regional Difficulties', Draft paper presented at the Dept. of Geography, Australian National University, 17th February.

Mullins, P. (1980) 'Australian Urbanisation and Queensland's Underdevelopment: A First Empirical Statement', International Journal of Urban and Regional Research, 4, 2:212-238.

Mullins, P. (1981) 'Theoretical Perspectives on Australian Urbanisation,1:Material Components in the Reproduction of Australian Labour Power', Australian and New Zealand Journal of Sociology, 17, 1:65-76.

Myrdal, G. (1957) Economic Theory and Underdeveloped Regions. London, Duckworth.

O'Connor, J. (1973) The Fiscal Crisis of the State. New York, St. Martin's Press.

Richards, J. and Pratt, L. (1979) Prairie Capitalism, Power and Influence in the New West. Toronto, McLelland and Stewart.

Pred, A. (1966) The Spatial Dynamics of United States Urban Industrial Growth, 1800-1894. The M.I.T. Press, Cambridge, Mass.

Saunders, P. (1981) Social Theory and the Urban Question. London, Hutchinson.

Schaffer, E. (1981) 'Class and Oil in Alberta', in Nore, P. and Turner, T. eds, Oil and Class Stuggle. London, Zed Press.

Stillwell, F.J.B. (1974) 'Economic Factors and the Growth of Cities', in Burnley, I.H. ed. Urbanization in Australia, the Postwar Experience. Cambridge, Cambridge University Press.

Wallerstein, I.M. (1974) The Modern World-System : capitalist agriculture and the origins of the European world-economy in the 16th century. New York, Academic Press.

Wilson, R.K. (1980) Australia's Resources and Their Development. Sydney, University of Sydney.

7 The city-bred child and urban reform in Melbourne 1900-1940

Graeme Davison

Somewhere between the mid-nineteenth and the early twentieth centuries there occurred a curious transformation in the image of Australian youth. The young colonial, as we meet him in fiction and social criticism during the early nineteenth century, is depicted as a tall and muscular lad, more robust than his convict or immigrant forebears and positively bursting with good health and animal spirits. Commissioner Bigge's famous description of the native youth - 'tall in person, and slender in their limbs...capable of undergoing more fatigue...than native Europeans...in their tempers...quick and erascible, but not vindictive...' is echoed by many later writers (quoted in Mcnab & Ward, 1962·291). By the 1870s and 1880s, as a new generation of goldrush children rose to their teens, a fresh theme emerged: there were fears that these well-fed youngsters, lacking the pioneering challenge of their fathers, would grow restless, bored and rebellious. R.E.N. Twopeny in his perceptive Town Life in Australia,(1883) noted that 'parental relations are undoubtedly the weak point of family life in the colonies. During childhood a certain obedience is of course enforced; but public feeling is strong in favour of the naughty boy and wilful girl, looking as it does upon these qualities as prophetic of future enterprise' (1883:101-2). There were complaints of the breakdown of the apprenticeship system, of the growing numbers of truants and 'street arabs', and of the unwillingness of colonial youths steadily to pursue their vocation (Davison, 1978:56-61; McConville, 1981:39-48). Around the 1890s a new and more sinister note was heard: increasingly the young Australian was pictured, not only as morally enfeebled, but as physically degenerate. Towards the end of the nineteenth century, if fiction is any guide, Australia's young men, particularly those who lived in the great cities, became thinner and shorter, turned sallow-faced and hollow-chested, and began to cough, smoke and spit. When he emerged, during the 1870s and 1880s, the larrikin was pictured as a rough and rebellious, but physically hearty, specimen. But in Edward Kinglake's portrait of 1891 he was 'generally a weedy youth, undersized and slight...He has a repulsive face, low forehead, small eyes, a colourless skin and irregular colourless teeth' (Kinglake, 1891:107). In his

'Captain of the Push', Henry Lawson (1892) described the 'beau-ideal of a Sydney larrikin' as 'bottle-shouldered, pale and thin' (1892:186-188). The heroes of Ambrose Pratt's and Edward Dyson's larrikin stories published shortly after the turn of the century were short, puny and sallow-faced. But the most extreme treatment of the degenerate type of city youth is found in Louis Stone's famous novel, Jonah (1910), in which the hero, Jonah, was actually a hunchback and the entire action was underscored with images of savagery and brutality.(1)

It is unlikely, of course, that the physique of Australia's city youth deteriorated quite as swiftly and completely as their literary reputation. Between the 1870s and early 1900s urban death rates steadily declined, real wages increased and so, probably, did general levels of nutrition and material well-being. The depression of the 1890s certainly reduced living standards but the decline in the young men's image seems to have preceded the depression and did not revive when it ended. In fact, the explanation does not lie in measurable changes in physique among the young so much as the changing outlook of their elders. The stunted, sallow larrikin was not a new fact of urban working-class life but a new figment of the urban middle class imagination.

To explain why intellectuals came to look upon the nation's youth with such misgivings we must enter the now-unfamiliar world of late nineteenth century eugenics. The stunted larrikin was the Australian variant of a social type which emerged, more or less simultaneously, in European and American cities and the disquiet of urban reformers in Sydney and Melbourne echoed contemporary debates about the city-bred child in London, Berlin and Chicago. The idea that city life was harmful to human health and morals was scarcely a new one, of course. Since the early nineteenth century sanitary reformers had asserted a close correspondence between the density or overcrowdedness of urban living space and the physical and moral health of the towndweller. Usually this was portrayed as a reciprocal relationship: bad conditions made bad people; bad people made bad conditions worse. But towards the end of the century a new and frightening factor entered these discussions. As the prudent middle class of the suburbs began to limit their families, the feckless people of the slums continued to reproduce as prolifically as ever. Medicos and sanitarians conjured up the dismal prospect that, over several generations, urbanization might create a permanently enfeebled and degenerate city race. The reproduction of unfavourable traits - stunted bodies, myopia, feeblemindedness - would render the towndwellers of the future less fit for life's struggle than their village forebears.

The 'theory of urban degeneration', as Stedman Jones (1971:Ch. 6) has aptly christened it, first appeared in a rudimentary form as early as the 1850s. 'The rapid increase of town populations', remarked the Bristol physician John Beddoe in 1857, was apparent in 'the low stature, the narrow chests and feeble frames of multitudes of our artisans and manufacturing operatives' (Beddoe,

1857:357; 1861:501-503). Viewed in the light of Charles Darwin's theory of natural selection, these symptoms of 'hereditary and constitutional weakness' were bad omens for the future of the city race. Darwinism was a versatile hypothesis and, as Hofstadter (1945:Chs. 8 and 9) has shown, it could be adapted to bolster a variety of optimistic and pessimistic, radical and reactionary social doctrines. Strictly interpreted, natural selection - unlike the Lamarckian theory of adaption - severely circumscribed the direct influence of environment upon the development of the species and implied a time-frame of millenia rather than mere decades. It did not assume, as urban degenerationists so readily did, that moral decline went hand-in-hand with physical decay. Nevertheless Darwinian concepts helped to give an air of scientific impartiality to an inner conviction among fin de siecle moralists that their civilization, as represented by the emerging generation of city youth, was in serious decline. The Maurice Committee established in 1902 to examine the physical fitness of military recruits to the Boer War found a disturbingly high proportion of rejects in the slum areas of the great cities (Gilbert, 1965; Searle, 1971:Chs. 2 and 4). As international tension mounted in Europe, public attention was increasingly focussed upon 'the boy life of our cities'. In his book The Town Child (1907) Reginald Bray, a London settlement worker, attempted to 'deduce the characteristics of the coming race, as it slowly emerges upon the plane of the world, by examining the reciprocal forces of environment stamping an influence on the race, and of the race struggling in mortal combat with the environment (as quoted in Reeder, 1977:79).

In the United States, where the test of war seemed more remote and urban degeneration less advanced, discussions of the city child were more optimistic, though no less intense, than in Britain. Progressive child reformers, such as Jane Addams, John Dewey and Louise de Koven Bowen aimed to forestall 'race suicide' by preventive and environmental measures rather than the harsh eugenic remedies advocated by some British reformers. The child-study movement founded in the 1890s by the psychologist G. Stanley Hall became the seed-bed for an entire repertoire of 'child-saving' reforms ranging from creches and kindergartens to municipal playgrounds, child-labour legislation and children's courts. 'If humanitarian progressivism had a central theme', writes Robert Wiebe, 'it was the child' (1967:169).

Definitions of the city-child problem reflected the circumstances of the countries in which it appeared. British historians have generally set debate about the 'boy problem' against the background of the struggle for mastery in Europe and have emphasized its eugenic and militaristic aspects. (see especially Searle, 1971 and 1976; Springhall 1977: Ch. 3; but compare Gillis 1974: Ch. 4). American historians, on the other hand, have delved more deeply into the tensions within their own society and have sought their explanations in those circumstances - the drift of population from farm to city, the influx of foreign-born, the prolongation of education and adolescent dependence - which inserted a sharper line of division between the experience of young people and their parents' generation (see Demos and Demos, 1969; and especially Kett 1977). Discussions of the city-bred child in Australia drew heavily upon both British and American sources although, as Carol Bacchi has persuasively argued, local commentators generally leaned towards the optimistic environmentalism of the Americans (Bacchi, 1980). The ways in which Australian thinkers drew upon, and adapted, overseas ideas are well illustrated in the work of two leading advocates of reform, C.E.W. Bean and James Barrett.

Two Prophets of Urban Degeneration: C.E.W. Bean and J.W. Barrett

Since Charles Bean was to become one of the most influential prophets of Australian bush nationalism, it is natural that historians should have emphasized his own Australian rural origins and experiences. Relatively little weight is given, by comparison, to the metropolitan intellectual influences which dominated the fifteen years between his departure from Sydney as a ten year old Bathurst schoolboy and his return as a young university graduate in 1904. Yet there can be little doubt of their formative importance. His public school in England, Clifton, was a kindergarten for young imperialists; Oxford, where he read classics and law, taught him to revere the Homeric ideals of physical strength, courage and sacrifice; while London, whence he had come most recently, exposed him to the theory, and disquieting evidence, of 'urban degeneration' (Inglis, 1979:226-229). In Australia, Bean soon forsook the law for journalism and in his first major series of articles, published in the Sydney Morning Herald in 1907, he applied the urgent lessons of 'Outcast London' to the young metropolises of Australia. Sydney, he conceded, had largely escaped the grinding poverty and gaping inequality of Old World cities. 'There is all the difference in the world ... between those who freeze and shiver the night through on the Victoria Embankment and the lotus-eaters who wrap a newspaper round their tired forms and sleep in the Sydney Domain', he claimed. But he was less hopeful about the physical and moral vigour of the Australian city dweller. 'The civilization of the Australian cities', he argued, 'is not new, but the old, hoary-headed decrepit European civilization which appears half the world over to be tottering to its grave'. The Australian cities might postpone, but could not forever escape, the fate of their European counterparts.

> The fate of ever city-bred people of any consequence that the world has known has been invariably the same, and unless there is something in the case of Australia which will differentiate it from all others, the fate of the Australian people will be in no wise different. As soon as a nation begins to shut itself up in cities, it begins to decay. First its bodily strength, and along with that its moral strength, declines. The new generations that have to bear the responsibilities of the old are smaller in body

> and weaker in courage and resolve than their forebears. The nerve of the nation, its level-headedness and fitness for a national emergency desert it as they are deserting the modern Londoner (Sydney Morning Herald, 8 June 1907).

Only if the environment of the Australian city was remade in a more healthful pattern could the threat of 'senile decay' be kept at bay. Bean called for a more ample provision of open space 'to preserve the country in the town' and for the promotion of vigorous manly sports among the city's youth. 'Properly encouraged, the Anglo-Saxon devotion to sport and athletics will save the Anglo-Saxon people', he prophecied. To rebuild Sydney on sound environmental principles would call for a comprehensive programme of street widening, slum reclamation, bridge-building, park construction and beautification (Sydney Morning Herald, 8th June, 1907). Like contemporary New Liberals in Britain, Bean recognized that efficient urban planning would need to overcome the fragmentation and self-interest of private enterprise and local municipal institutions, and throughout his life he continued to preach the importance of town planning and of a 'Greater Sydney' scheme (see especially Bean 1919: Ch. 13; 1943: Ch. 13).

Between Bean's early efforts as an urban reformer and his later, and more famous, career as Australia's official war historian there was an essential continuity of concern for the future of the Anglo-Saxon race. Indeed, his 'New Chum's' fears for Australia's city-bred youth may have been as important as his later experiences in outback New South Wales or, even on the slopes of Gallipoli, in forming his ideal of the Australian bushman soldier. The tall, clear-eyed, muscular bushman was conceived as a foil for the myopic dwarf-men of the great cities. Even before he had himself set out 'on the wooltrack', when his most recent acquaintance with rural Australians had apparently been through jingoistic stories of their military prowess during the Boer War and from observation of 'Australian boys at English (public) schools', Bean had formed a clear-cut idea of their virtues. The courage, candour and comradeship which he later found among the Anzacs were already shown by the hero of his unpublished Boer War story, 'Bill's Brother', written around 1905. According to its narrator 'who was never at the war', the details of the story came from a 'friend outback'. But the qualities of his hero - 'an eye for the country and a way to treat a man as a man' - were those of a generalized frontier type rather than a recognizable portrait of an Australian bushman (Bean Papers, n.d.).

Bean was not alone in his admiration for the soldierly qualities of the frontiersman. The idea of the great outdoors as the natural training ground for fighting men was widespread in Britain's dominions during the Edwardian era. When war came politicians and recruiting officers used it to lure the sometimes reluctant farm lads to the Front, and when hostilities were over historians enshrined it in national tradition. But whenever the ideal of the frontier soldier appeared, the contrasting image of the unfit city type was never far away. Bean's celebrated account of the Australian Imperial Force, in the opening chapters of his volumes on 'The Story Of Anzac', portray the Australian soldier as a result of the 'evolution' of the 'Anglo-Saxon stock' under the special environmental conditions of the Australian outback (Bean, 1937, 1:45-47, 126).(2) Throughout his Official History, the physical stature, personal resourcefulness and mateship of the Australian bushman are contrasted with the puny physique and moral servility of the city-bred British 'Tommy'. Likewise, New Zealand's war historian exalted the 'martial instincts' and personal independence of those farmer-soldiers who had grown up 'far from the baneful influence of the great cities' (Waite, 1921:1). As the Australian outback had bred a special brand of physical courage, so - it was argued - had the Canadian backwoodsman's 'war with Nature and with wild beasts' bred qualities of 'decision, initiative, resource, endless courage and capacity for endurance such as one hardly found in the men of towns and cities' (de Beck 1917, 1:36). Throughout the Empire, thin, bespectacled men in city offices joined their praises of frontier manhood and the character-building rigours of the Great Outdoors.

Bean's viewpoint, as I have tried to show, was that of a classicist and a literary man and his vision of urban decadence, like his ideal of manhood, owed as much to ancient Greece and modern London as it did to his observations of Sydney or the Bush. James Barrett, on the other hand, was a scientist and a professional man whose anxieties about the future of Australia's urban youth found expression primarily in scientific investigation and practical reform. The son of a doctor who had migrated to Victoria in the 1850s, Barrett was himself among the stars of the remarkable second generation of native-born medicos who graduated from the University of Melbourne in the 1880s. He served briefly as a resident of the Melbourne Hospital and wrote a pamphlet on the city's typhoid problem before undertaking specialist studies in ophthalmology in London. Barrett's 'able, active and thrusting personality' - the description is Bean's - quickly brought him to the forefront of public life in Melbourne. From the time of his return in 1887 he became an activist in the fledgling British Medical Association; during the 1890s he was elected to the University Council; and throughout the early twentieth century he was notoriously the city's busiest committeeman and most tireless writer of letters to the editor. At the height of his influence between 1910 and 1940 he was the leading spokesman and organizational pivot of progressive reform in Melbourne, occupying key positions in such diverse fields as town-planning, slum abolition, child health, workers' education, nursing, kindergartens and national parks (Murray-Smith, 1979; Obituary in Australian Medical Journal, 1945).

But if Barrett's interests were many, his ideals were singularly unified and compelling. Like many other Australian participants in the 'nature - nurture debate', Barrett attempted to straddle the middle ground between the radical eugenists with their stress upon the elimination of the 'unfit' and the ameliorative environmentalists who placed their faith in practical schemes of urban reform. He shared the hereditarians' concern that the decline of the middle-class birth rate would depress the quality of the race. As a doctor he had at first welcomed the spread of contraceptive knowledge as a means of rearing 'better children' with fewer deaths but the rapid decline of the birthrate in the 1890s convinced him that birth control had become 'an instrument of national destruction' and after 1900 he was prominent among those moralizing medicos who urged the educated middle classes to do their procreative duty (Hicks, 1978:Ch. 3). The hereditarian strain in Barrett's thinking, however, was offset by his belief in the modifying effects of environment. Like many of his contemporaries he was vague about the precise relationship between the growth of cities and the deterioration of the race. Sometimes, it appears, they emphasized the effects of overcrowding and poverty, so that moral deterioration went hand in hand with physical. But elswhere it was the luxury and artificiality of city life, heightened perhaps by its abnormal pace and anxiety, which they saw as the destroyers of the emerging city race. In accordance with a popular interpretation of Darwin's theory of natural selection, Barrett believed that human physique and character were improved by the struggle against physical adversity and hardship:

> Character ... is in all probability the direct product of the action of environment on plastic living matter. Just as the physical characteristics are the direct produce of the strains and stresses to which it has been subject for aeons of time, so the expression of the conscious nervous system is in all probability an indirect product of environment, that environment meaning not only the physical factors which are influencing us, but also the super-organic factors of social and commercial life, which are also playing their part (Barrett, n.d. 'National Character').

A child bought up among criminals was therefore likely to become a criminal himself. A more envigorating climate might make labourers more hard-working. And while a boom, like that of the 1880s, might have a demoralizing effect, the succeeding depression would act 'very much as a cold spinal douche on the community' in sobering it up. The ultimate test of the race, of course, was war. 'If adversity is the essential factor in the production of sterling character, Victoria', Barrett concluded, '... has had a proper proportion in quantity; the only doubt remains whether she has had the right quality; that is to say, whether the Victorian character will be complete until a few tens of thousands of native-born have been mown down with Maxims and 4.7 inch guns' (Barrett n.d. 'National Character'). This awesome prophecy, uttered almost a decade before Gallipoli, gives us a sudden insight into the powerful psychological urge to war that was building up among the nation's leaders during the Edwardian era and foreshadows Barrett's own active role in the campaign for compulsory military training and during the Great War itself when his own son was among the young Victorians 'mown down'.

If 'character', as Barrett asserted, was the product of adversity, society's natural leaders were those who had proved their fitness in the struggle of life. Barrett's political ideology departed from orthodox doctrines of either the Right or Left for he divided society, not between 'labour' and 'capital', but along meritocratic lines between the 'productive' and 'unproductive'. Among the 'producers' - a category similar to Saint-Simon's *industriels* - he numbered all those who contributed by hand or brain to the common good. The 'non-producers', on the other hand, included members of both the highest strata - such as stock exchange speculators and rentiers - and of the lowest - such as recipients of charity. Inherited riches and chronic pauperism were alike in their effects upon character since they both shielded the individual from the testing struggle for survival. Among the producers, however, some were better equipped by education for the tasks of leadership. The education of manual and clerical workers, with its emphasis upon rote-knowledge and practical skills, was inappropriate for the higher duties of domestic and political life. But medical men, by virtue of their knowledge of biology and scientific method, commanded the key to political science. Barrett commented,

> Members of the medical profession ... receive an exceedingly long drawn out technical education which ... fits them for something else besides practice - it gives them a sound logical basis of scientific information on which they can build if they will the most complex mental superstructure resting on the firmest and most solid of foundations, those laid by natural science ... On such a basis alone can anyone hope to gain a comprehensive knowledge of the significance and importance of the facts manifested in our daily personal and social life (Barrett, n.d . 'Education').

Other professions - engineers, lawyers, musicians - shared some of these qualities but medical men, who understood the mysteries of life, were preeminently qualified for leadership. Barrett therefore belonged to that line of nineteenth and twentieth century technocrats, beginning with Saint-Simon and Comte, who wished to reorganise society on scientific lines, and who looked to a new caste of experts - 'an aristocracy of intellect', as he called it - to bring this about.

The Class Character of Reform

In Melbourne, Barrett became the spokesman for a group of professional middle class reformers whose ideology and way of life clearly marked them off from both the propertied or business class and from the organized working class. They were critical of the egalitarian aspirations of the Labor Party, which they equated with the '"Something for Nothing" attitude', and their formal political connections were with the Deakinite element of the Liberal Party (Barrett, n.d. 'Socialism'). Yet they were equally opposed to the unrestrained individualism of the mercantile bourgeoisie and argued that the socialist use of state power was not only necessary but inevitable. They had much in common, socially and ideologically, with the American Progressives and the British 'New Liberals' from whom they borrowed many of their practical schemes of reform. Their influence was centred in the professional faculties of the University of Melbourne and extended into the learned professions, especially medicine and law, and through the networks of organised philantrophy. In the institutions of elite reform, as in the professional circles which they so closely reproduced, family connections and traditions played a significant part and on the boards and committees of the main reform organizations the names of prominent professional families, such as the Brookeses, A'Becketts, Deakins, Derhams and Norrises, frequently appeared. So far as this network had a residential base it lay in the old elite areas of South Yarra and St. Kilda, although many of its younger members were spread through the respectable railway suburbs to the south and east of the city. Compared with their late nineteenth century predecessors - the generation of Deakin, Pearson, Sutherland and Strong - they were both more scientific and more secular in outlook, although not necessarily less imbued with a sense of vocation. 'In the gospel of Social Service', Barrett confessed in a revealing passage, 'many fine minds find solace and relief; for men thrown back on themselves by reason of the difficulty or impossibility of obtaining what appears to them an adequate solution to the problem of their relationship to the universe, turn in the direction of practical idealism, and find their happiness and interest in attempting to unravel numerous important and complicated social problems' (Barrett, 1918, 1:218).

The aims and methods of the reform movement were probably shaped as much by the subjective concerns of its professional middle-class constituents as they were by the objective needs of its working-class clients. The high social prestige which doctors and lawyers had recently won for theselves often carried with it a nagging sense of insecurity. In spite of their standing they could not anticipate, as confidently as those who owned landed or industrial property, that their children could also stand as high. To maintain their family's position in society the new generation would have to climb the same difficult route. The costs of professional qualification may have made professional men and their wives more sensitive to the burdens of parenthood. Certainly they, along with the rest of the 'new' middle class, appear to have led the general movement towards family limitation which swept through Australian cities in the depressed 1890s.(3) According to historians of the United States, where similar trends unfolded, these changes in the structure of the middle class family brought subtle, but significant, changes in its emotional climate. Limiting the number of children in the family increased the psychological investment of parents in each individual child. Prolonging the period of preparation for a career, it is argued, increased the sense of financial and emotional dependence among adolescent children themselves (Platt, 1969:98-9; Kett, 1977:144-5,235). Until more is known about demographic and social trends in early twentieth century employment and education, it will be difficult to say how far Australian middle-class families followed the American pattern. We know that the early twentieth century saw a substantial increase in the proportion of adolescents, especially males, continuing in full-time education and in the proportion of children and adolescents being educated in private schools and universities (Victorian Yearbook, 1911-12:517, 520). In Melbourne, dependent adolescents, private school pupils and university students were concentrated in the newer south-eastern 'Bible-Belt' suburbs, such as Hawthorn, Kew, Camberwell, Caulfield, Malvern and Brighton.(4) The idea of adolescence as a period of financial dependence and emotional turmoil, which was popularized in the United States by G. Stanley Hall in his famous book Adolescence (1907), also won early acceptance among Australian child experts.(5) So it is possible that the sense of moral crisis which pervaded public discussions of slum youth drew in part upon the vocational dilemmas of the professional middle-class family. The prospects of slum children and middle-class adolescents were also linked, of course, in a much more direct fashion, for by providing playgrounds, kindergartens and baby-health centres to improve the first the reformers greatly expanded the field of respectable employment for the second. By highlighting the physical and mental problems of the city-bred child the professions also helped to create the need for a new range of professional or quasi-professional jobs for their sons, and especially their daughters, as kindergarten teachers, child welfare nurses, play leaders and so on.

By placing the urban reform movement in its proper urban and class context, as an expression of the distinctive ideology of an emergent professional middle-class, we are better able to understand the course of its development and the limitations upon its success. Recent historians of town planning and public housing, such as Sandercock (1975) and Spearritt (1974), view these early movements from the assumed vantage point of a socialist planner and tell their story as a contest between altruistic, but befuddled, planning theorists on the one hand and rapacious businessmen and corrupt politicians on the other. 'The City Beautiful', according to this view, was strangled at birth by big business. Planners themselves succumbed to the prevailing laissez-faire ideology for 'their reformist intentions', we are told, 'were limited to palliatives that would strengthen rather than threaten the status quo' (Sandercock, 1975:25). The trouble with these histories is that they fail to reveal the essential objective and inner logic of the reformers' programme. They were neither simple humanitarians nor would-be comprehensive planners who were somehow deflected from their goal by commercial or political pressures, but social scientists concerned above all for the fitness of the future Australian race. In their belief, the most fundamental questions were not about wickedness in the human heart nor the evils of unbridled capitalism, but the physical and moral degeneration of the nation's youth. We may agree that theirs was an ideological approach to urban problems - a distorted view based upon class interest - but it is important, if the movement is to be interpreted correctly, to recognize its intellectual coherence and the distinctive professional-class interest which it expressed. Otherwise it becomes difficult to explain why, when they were confronted with the failure of simple environmental solutions to urban problems, the reformers did not tackle their (to us) 'basic' economic causes but directed their attention backwards to earlier, and as they thought more fundamental, stages of the human life cycle. Over the first decades of the twentieth century the progressive reformers came to adopt an increasingly pessimistic view of the Australian city race and shifted the focus of their efforts backwards from adolescence, through childhood to the conditions of procreation and childbirth. In a broad way the peaks of reforming activity may be linked to phases of the human life cycle so that while the first decade of the century may be thought of as 'the age of adolescence' and the second 'the age of the child', the 1920s are marked out as 'the age of the infant'. Measured against the methods and objectives of the modern urban planner, the progressive reformers may be judged to have failed. But in the first quarter of the twentieth century they established some of the most familiar features of the Australian suburban landscape: the baby health centre, the kindergarten, the municipal playground, the Scout Hall and the High School.

The Age of Adolescence

When we consider how widespread and vociferous were the complaints about the degeneration of urban youth in the decade before the Great War it is remarkable how little hard evidence of their delinquency can be found. Politicians and preachers declaimed about 'the growing evil [of] contamination of boys by street life' but in spite of increased vigilence on the part of the police, the level of convictions among juveniles for 'larrikin' offences (abusive language, assault etc) remained fairly constant throughout the first decade or so of the century.(6) There was, admittedly, a substantial, but uneven, growth in the number of children being brought under the care of the Victorian Neglected Children's Department but these fluctuations were plausibly explained by the officials themselves in terms of changing criteria for admission, and increased hardship among parents, rather than growth of the number of uncontrollable young people (Reports of the Department of

Neglected Children, various years).

All the more remarkable, therefore, that this period should have witnessed the establishment of so many new institutions directed specifically at the reformation of city youth. Common to all this legislation was the idea that young people between the ages of 12 and 18 were subject to special physical and moral dangers and in need of specific preventive care and discipline. The main instrument of reform, of course, was the State School and in 1910 the compulsory clauses of the Education Act were strengthened to require the regular attendance of children up to the age of 14 (Commonwealth Yearbook, 1917:812-3). Within the school itself there was a new stress upon the physical training in the form of posture and drill and upon schemes of vocational and technical training that would make good the deficiencies of the apprenticeship system and save working class lads from dead-end jobs (Commonwealth Yearbook, 1917:812-3; Barrett, 1918, Vol. 1:271-2; on posture see Roth, 1914:774-7). The education of girls was also reformed on eugenic principles, with heavier emphasis upon mothercraft and domestic science to ensure their 'complete fitness for maternity and the practical care of a home' (Roth, 1914:42-4; and compare King, 1914:83-4). In 1910, the Minister of Public Instruction reported that the schools had already 'done much to counteract the decline of household skill due to the changing conditions of modern city life' (Report of the Minister of Public Instruction, 1912:20). The establishment during the same year of the first government High Schools was a further step towards the institutionalization of adolescence as a time of preparation for an adult vocation (Austin and Selleck, 1975:321-40).

Outside the school adolescents were also subjected to increased srutiny and discipline. Medicos, appalled by the 'dwarfed constitutions, broken down nervous systems [and] sallow complexions' caused by smoking, secured legislation in three states outlawing the sale of tobacco products to juveniles and prohibiting the youngsters themselves from smoking, at least in public (Dunstan, 1974:136-7). The Victorian Children's Court, established in 1906 on the pattern of the famous Illinois prototype, put emphasis upon the rehabilitation rather than the punishment of young offenders, a shift which may nevertheless have intensified, instead of relaxing, their custodial treatment (see Victorian Parliamentary Debates; and compare Platt, 1969:145). The Edwardian period also saw an upsurge in youth organizations of a highly disciplined and often paramilitary type. The most popular of these, the Boy Scouts, was a direct product of the Boer War scare about the fitness of British urban youth for its founder, the hero of Mafeking, Lieut.-General Baden-Powell, had invented scouting as a means of exposing lads from the London slums to the character-building influence of the Great Outdoors. Scarcely a year after his famous experiment at Brownsea Island the Boy Scouts had been launched in Australia and by the end of 1910 they had enrolled their first 3,000 'tenderfeet' in Melbourne (Argus, 1st October, 1910; 6th January,

1911). Scouting generally attracted those who favoured a militaristic rather than a more subtle environmentalist approach to juvenile reform and within a year or so of its foundation in Melbourne the movement was split between the quiet outdoors men who leaned towards bushcraft and hiking and the tougher types, like John Monash, who stressed marching and drill (Argus, 3rd February, 8th February, 1911).(7) It is ironic that when Baden-Powell visited Australia in 1912 he was more impressed with the salubrity than the squalor of the Australian cities. 'What a place for Scouting!' he exclaimed on reaching Sydney. 'All around the town and even in it are the wooded hills and the thick bush; while the harbour with its islands and creeks and innumerable boats gives the finest field for sea-scouting I have ever seen. Lucky beggars those Sydney boys!' (Baden-Powell, 1913:143). His local supporters however, increasingly saw scouting as an introduction for younger adolescent boys to the compulsory military training introduced by the Commonwealth government under its Compulsory Service Act of 1909. J.W. Barrett, himself an enthusiastic volunteer soldier, emphasized the eugenic rationale for the scheme:

> There is too abundant evidence that in the pursuit of trade or profession, or other forms of city indoor life, there is a definite tendency to physical deterioration. The man in the street knows that this is so. The physiologist accurately measures the changes, and possesses evidence that it is a tendency that must be met. [Barrett had himself compared the incidence of myopia among country and city subjects, to the disadvantage of the latter.] The human being has through countless centuries adapted himself to the outdoor life in various climates. He cannot in a century or two adapt himself to the high pressure of modern industry, without the physical organism under-going considerable alteration characterized by depreciation of tone ... (Barrett, 1918, Vol. 2:259).

No doubt it could be argued, as Baden-Powell and Bean had done, that the relatively favourable conditions of Australian urban life, with its abundant facilities for manly outdoor sports, would sufficiently counteract these tendencies. But Barrett (accurately, we might now admit) foresaw a future Australian race of flabby spectators rather than a nation of bronzed athletes. Only a due measure of compulsion, he argued, would get the larrikin off his backside and into a drill hall.

So it proved. Two recent studies of the working of the Compulsory Training Act have offered contrasting verdicts on the support which it received among Australian city youth. T.W. Tanner following in the anti-conscriptionist tradition, emphasizes the ideological opposition to the scheme and cites the more than 27,000 prosecutions of boys who failed to comply with its requirements. While acknowledging that most of these delinquents objected more to the inconvenience than to the principle of compusory service, he estimates that the level of non-compliance was about fourteen times as great, in relation to the eligible population, as it was under the compulsory clauses of the Education Acts (Tanner, 1980:180-186, 192, 202-3, 205). John Barrett (1979), on the other hand, contends that all but a few of the boys liked, or at least did not object to, their stint of soldiering and that the scheme was generally supported by the public. He backs his view with an impressive array of archival and interview evidence. Yet the difference between these two studies is as much a matter of perspective as evidence: the number of non-compliers may have been small in relation to the very large population of eligible boys while nevertheless creating a very sizeable problem of law-enforcement for police and magistrates in certain areas. An act which commandeered youthful leisure time for strenuous manoevres on the parade-ground bore most heavily upon those with least leisure and energy left over from work or school and it is not surprising perhaps that levels of non-compliance appear to have been greatest in working-class districts (Barrett, 1979, Chs. 5 and 6). Magistrates in the inner suburbs of Melbourne were kept busy dealing with those who refused, or neglected, to register or attend parades and it may be significant that, at a time when the level of arrests was declining throughout the metropolis, they continued to increase in working class Collingwood (McConville, 1974:104, 291).

Indeed implementation of the Compulsory Service Act only confirmed the diagnosis of city youth which had inspired it. Following the usage of contemporaries, John Barrett describes the non-compliers as coming from among 'the idle and insolent', 'the larrikins and roughs', 'the lazy and the lout' (Barrett, 1979:185, 187, 241, 244). The truth of course was that the more reformers focussed on the condition of the adolescent, the more agencies they established to educate or discipline him, the more inquiries they conducted into his physique or morals,then the more, paradoxically, his condition must appear to deteriorate. To increase the occasions for compliance was necessarily to increase the level of apparent deviance. As the range of investigations was broadened, new defects were brought into view. So, for example, in 1911 when the Commonwealth Statistician, George Knibbs, carried out the first systematic comparison of census and school attendance figures he revealed that in Sydney and Melbourne no fewer than 10,000 school-age children were not attending school (Commonwealth Census, Statistician's Report, 1911:196-7). Similarly, in 1910, when the Victorian Education Department carried out a preliminary medical inspection, its officers brought to light 'the existence of a vast amount of hitherto unrecognized physical defect and disease amongst the children of our schools' (Reports of Minister for Public Instruction, 1910 and 1911). Not only were city children less fit than their country counterparts, but those who lived in the crowded industrial suburbs were smaller and less robust than the children in the outer suburbs. Thirteen-year-old boys from the

Essendon State School, for example, were, on average, 1.5 inches taller and 10.5 pounds heavier than those from Vere Street, Collingwood. Slum children were more prone to have rickets, bad teeth, chest and nasal infections and were more likely to be intellectually backward and morally delinquent. According to the officer in charge of the investigation, Dr. Harvey Sutton, the association between environment and health was simply one of 'cause and effect: that is, the crowded accommodation is directly responsible for the interference with growth' (Royal Commission on the Housing of the People in the Metropolis, 1917, Qs 4973-80).(8)

The Age of the Child

In response to these sobering discoveries, and in order to break the vicious circle of environmental contamination, the reformers gradually shifted the focus of their concern backward from the adolescent to the child and outward from the young delinquent himself to his physical environment. The Education Department's medical officers had themselves detected,

> a growing conviction that to secure a solution of the many grave problems of disease and degeneration in our community, the best and most economical method is not to wait till the results have developed and then provide for them in hospitals, asylums, goals, and charities generally but to attack these problems at the earliest possible moment ... While it is not at present permissable to control parentage, the health and development of the human product and the control of his environment are matters very much at our disposal. It is felt that every child should have full opportunity to develop the best in him, and thus become a more efficient citizen. ...After all, the men that are made in Australia must be the main bulwark of our Commonwealth defence, and no better method could be devised of furthering this end than by developing physically efficient boys and girls, the future citizens of our land (Report of the Minister for Public Instruction, 1910:96-97).

As they swung towards a more environmentalist approach Australian urban reformers came more directly under the influence of American ideas. J.W. Barrett's visit to the United States in 1912 coincided with the peak of the child-saving movement among the Progressives. 'The day of the child has come', an enthusiastic reformer proclaimed at a great conference on 'The Child and the City' held in Chicago during that year (Breckinridge, 1912:6). Barrett visited settlement houses, playgrounds and university departments concerned with the child problem in Chicago and on his return to Melbourne lantern slides of his visit and references to American experiments in child reform became a feature of his tireless public speaking (Barrett Papers, nd:Child Welfare). One of his main publicity *coups* was his stage-management of the Select Committee on the Housing of the People in the Metropolis which he used as a platform for his ideas on town-planning, building regulations, municipal playgrounds and pure milk supplies for infants. Melbourne's slums, he readily conceded, were 'not so objectionable as slums in older cities. My interest in the matter takes the form of an endeavour to prevent those conditions developing in a young country which have developed in old countries...' (Select Committee on the Housing of the People in the Metropolis, 1913-1914:Q368). Among the doctors, social workers, policemen and clergymen who appeared before the inquiry - many of them, one suspects, at Barrett's instigation - there was a consensus about the importance and practicality of reforming the city child's environment. 'Do you think the locality has anything to do with the general upbringing of the child?' Deaconess Henderson was asked. She replied: 'I think mothers and fathers are tremendously handicapped in trying to bring up children in these lanes, because of the contaminating nature of the surroundings and the danger to health in small ill-ventilated rooms. Their health must suffer, and health affects the moral character' (Royal Commission on the Housing of the People in the Metropolis, 1917:Q 2126).

A comprehensive answer to the problems of the city-bred child called for an ambitious programme of slum clearance and the rehousing of the slum dwellers in airy garden suburbs (Select Committee on the Housing of the People in the Metropolis, 1913-14:Q 368-373; Barrett, 1918, Vol. 2:97-98 and Sandercock, 1975:64ff). Meanwhile, however, reformers believed that much could be achieved by planting little oases of health and enlightenment, in the form of playgrounds and kindergartens, in the midst of the most blighted neighbourhoods. A model playground, on the American pattern, had been shown by the National Council of Women at the Australian Women's Work Exhibition in 1907, and in the following year an equipped, but unsupervised, playground was opened on land made available by the Melbourne City Council in Lincoln Square, Carlton (Norris, 1978:19-20). After a promising start the movement then appears to have faltered and it was not until the return of Miss Madeline Murray of the Victorian Neglected Children's Department from a tour of the United States in 1912 that public interest was rekindled. A small group of enthusiasts in the Liberal Party, led by Mrs. Herbert Brookes and under the patronage of her father, Alfred Deakin, established a Guild of Play and recruited volunteer supervisors for the Lincoln Square playground (Playgrounds Association of Victoria, nd). Ivy Brookes' manifesto for the Guild reflected the democratic and environmentalist assumptions of the contemporary American playground movement.

> Life [she wrote] is 'responsive to environment.' If the environment is only a fence, a lamp post, a rubbish tin, a hawker's cart, a passing motor, the response in too many cases may be predicted. If the fence draws attention it is

climbed or pulled down; the glass in the lamp-post becomes a target; the rubbish tin invites a kick, the hawker's barrow a possible theft. What is the life-character-personality of the child if his expressions of life or responses are limited to such environments?' (Brookes, n.d.).

The question, of course, was rhetorical for there was a consensus among the child reformers that the urban environment had a determining influence upon those who were reared in it. The Guild of Play, and its successor the Playgrounds Association of Victoria, aimed through the use of 'free play' to counteract the harmful effects of the slum environment. American playground advocates were divided upon the question of how structured and supervised the child's play should be. Deweyan democracy said that children learned simply by being exposed to a free and natural environment; but middle class common sense dictated that he be firmly led along the right path. The announced objective of the Guild of Play - 'to make free play unconsciously purposeful' - reflected this dilemma but also managed to convey the disciplinary purpose which always lay behind the liberal rhetoric of the reform movement (Brookes, nd; and compare Dewey, 1916; Gulick, 1907:481-86; Veiller 1907:509).

A similar tension between democratic theory and authoritarian practice pervaded the work of the free kindergarten movement. From the time of its foundation in 1908, by a group of middle class ladies closely associated with the National Council of Women, the Free Kindergarten Union rapidly expanded its activities and by 1911, when a galaxy of Victorian Liberal politicians attended a special display of kindergarten work in the Melbourne Town Hall, it had planted ten free kindergartens in the 'congested areas' of the city (Argus, 5th and 30th March, 1911; A'Beckett, 1935-36:16-17). Many of the early kindergartens had been strongly influenced by the progressive educational theories of Froebel, Montessori and Dewey and some of those who taught at the Kindergarten Training College, established in 1918, had themselves absorbed progressive ideals at their source in the United States. 'We found out that those who came from the American kindergaren colleges in the inspiring days of our studentship brought the new education with them', an old student recalled. 'From this very centre itself in Chicago, and on fire with enthusiasm came to Australia those American kindergarteners under whom some of us were privileged to train in the early days of the movement' (Fawcett, 1929:29). The usual kindergarten programme - the familiar routine of 'Milk and Fruit', morning sleep, spontaneous outdoor play and nature study - and the architecture of the kindergarten itself, with its kitchen, bathroom, doll's corner and well-ventilated sleeping area, were consciously designed to repair the deficiencies of the slum environment (Free Kindergarten Union, 1925-26:13; 1930-31:5).(9) Yet the 'new' educational ideals of spontaneity and naturalism were always balanced against traditional notions of order and discipline. The objective, as the Free Kindergarten Union's

MAP 1 Reform Institutions in South Carlton c.1914.

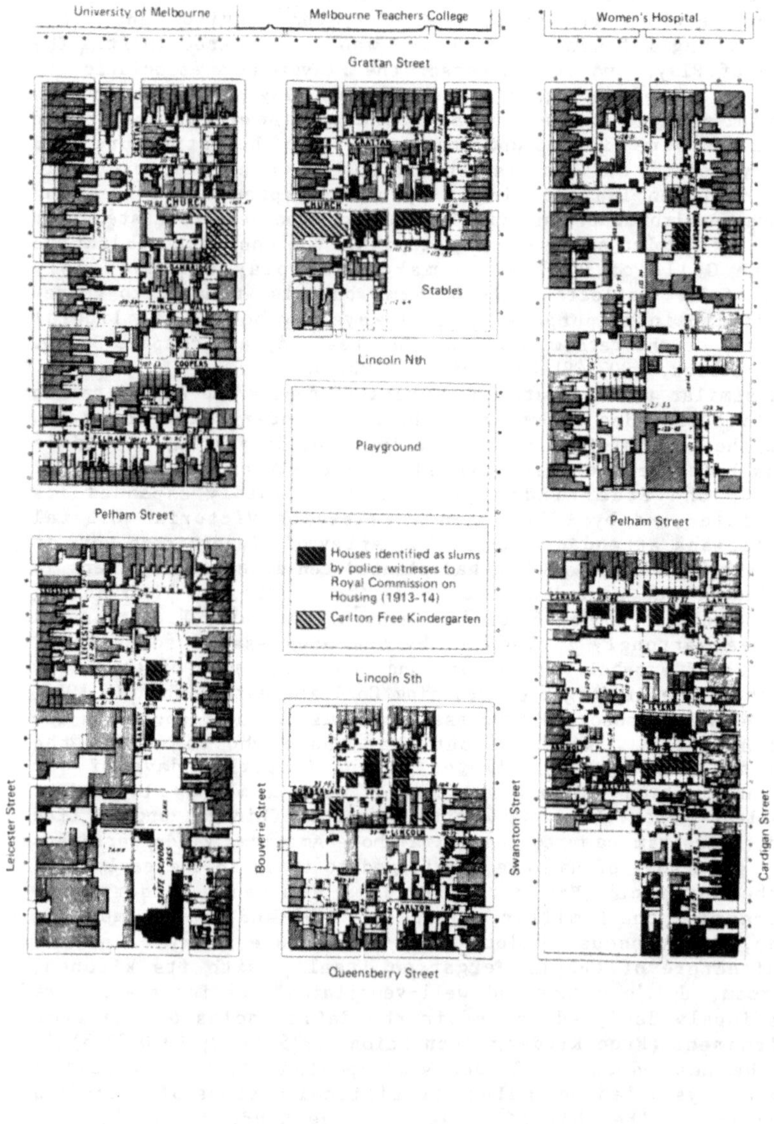

long-time president Ada A'Beckett explained, was 'to know how to lead [the deprived child] to do right, how to encourage him to think for himself along right lines of law and order, how to manage for himself so as to become a useful citizen.' (Free Kindergarten Union, 1934-35:18).

By the early 1920s the progressive reformers had encircled the inner city with a veritable 'cordon of kindergartens' (10). By 1923 there were supervised playgrounds operating in Carlton, South Melbourne, Collingwood and the Flagstaff Gardens and no fewer than 28 free kindergartens and nurseries, most of them concentrated in the inner working class suburbs, and especially in the notorious slum pockets identified by police witnesses before the Royal Commission on Housing in such areas as the east end of Little Lonsdale street, South Collingwood, Burnley, South Fitzroy, Montague and South Carlton (Map 1).

MAP 2 Kindergartens and Kindergarteners in Melbourne c.1928.

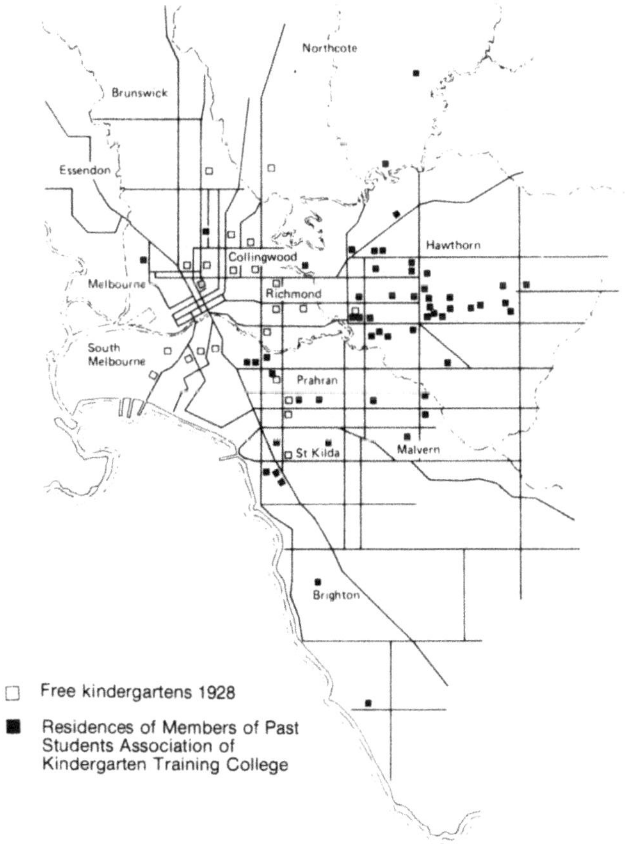

The ecological assumptions of the reform movement are clearly apparent in the social landscape of Lincoln Square, South Carlton where the professional training schools of the University, the Melbourne Teachers' College and the Women's Hospital stood on the very doorstep of an area notorious for its depressed piecework industry, decayed housing and endemic lawlessness. (The 'Bouverie Street Push', recruited in this neighbourhood, had been one of the city's most notorious larrikin gangs) (McConville, 1974:70-137; Select Committee on the Housing of the People of the Metropolis, 1913-1914:Q 1324). The playground and kindergarten planted in the midst of this area thus performed a dual role as agencies of child reform and professional training grounds.

The middle-aged matrons who founded and financed the free kindergartens and the young women who became kindergarteners and playleaders were drawn overwhelmingly from the middle class suburbs to the south-east. Ladies in the elite suburbs of Kew, Hawthorn and Toorak 'adopted' deprived children in the adjoining working class areas of Collingwood, Richmond and South Melbourne. The younger generation of kindergarten teachers, however, seem to have come more often from the newer 'Bible-Belt' suburbs of Malvern, Camberwell and Caulfield (Map 2). (Kindergarten Magazine, 1929(5):43-4). Kindergartening, as a contemporary politician remarked, was 'essentially women's work for which they are well fitted' (J. Mackinnon, Argus 25th October, 1910) and, as Spearritt has shown for the Sydney movement, it offered an attractive interim vocation for well-educated young ladies destined ultimately for motherhood (Spearritt, 1974:185-6). Mrs. George Swinburne, wife of a Liberal Politician and mother of a playleader, observed how often it happened 'that when the schooldays were finished girls did not know how to fill in their time. Through the agency of the "playschool" girls not only found occupation, but they also obtained a rational understanding of child nature' (Argus, 28th April, 1911). The prospectus of the Kindergarten Training College emphasized that kindergartening, 'unlike other forms of commercial and professional work which are open to women, fosters and develops all the tastes and interests that go to the making of an efficient and happy home-maker' (Kindergarten Training College, 1926-27:8). As a career, kindergartening offered more psychological satisfactions than material rewards and, since idealistic graduates were all too swiftly transformed into 'efficient and happy home-makers', the supply of teachers seldom outstripped the needs of the twenty or so inner city free kindergartens.

The Age of the Infant

By about 1920, in fact, the movement to reform the environment of the child had almost run its course. Neither the playgrounds nor the free kindergarten movement grew substantially after about 1920 and the liberal reform movement, now increasingly a professional rather than a voluntary affair, turned its attention to the conditions affecting the healthy nurture of small babies.(11) If the British eugenists had dominated discussion in the first decade of the century, and American environmentalists in the years around the Great War, in the 1920s it was particularly the example of New Zealand and the influence of Dr. (later Sir) Truby King, the famous mother and baby care expert, which came to the fore.(12) The salient features of King's approach - regular breast feeding, scrupulous sterilization, regular weighing and measuring - represented an extension of those disciplines which advocates of military training and scouting had once imposed upon adolescents, and kindergartens had sought to inculate among slum children, to the nurture of tiny babies. The ideas which King had popularised in New Zealand through the Plunkett Society were already current among some Melbourne reformers, including Barrett, before the Great War. In line with his general approach, Barrett saw baby care as a measure of 'practical eugenics'. 'It was really an attempt to secure a strong and healthy race by constructive, and not restrictive, means' (Barrett, nd: 'Child Welfare'). In Sydney, Dr. W.G. Armstrong had already inaugurated a network of infant welfare centres throughout the inner suburbs and by 1914, when the scheme was taken over by the Department of Public Health, his team of nurses had visited 28,000 newborn babies and their mothers (Armstrong, 1939:672). In Melbourne, the movement was slower to get under way and it was only in the aftermath of the Great War, when mass slaughter had revived fears of national depopulation, that the first infant welfare centres were established. A group of voluntary workers led by Dr. Isabel Younger had began work in the inner suburbs during the closing years of the war and in 1918 they formally inaugurated the Victorian Baby Health Centres Association (Scantlebury & Main, 1926:12-36). In the following year Dr. Vera Scantlebury was appointed honorary medical officer to the Association. Daughter of a medical family and an able graduate of the University of Melbourne, she had returned from service with the Royal Army Medical Corps in London only to find the profession overstocked with repatriated army doctors. She turned to infant welfare work almost accidently, as a useful interim job rather than a lifetime vocation. 'It is sensible work, approaching matters from a prophylactic standpoint', she explained to her mother, although she emphasized that it was of advantage to her career only as an 'advertisement' (as quoted in Kapper, 1977:3-5; see also obituary in Medical Journal of Australia, 1946:393-5). Her reputation grew with the movement and in 1923 she was one of two women doctors sent by the Victorian government to investigate maternal and infant welfare in New Zealand and Canada (Scantlebury, 1927a, 1928). Upon her return she was appointed director of the Infant and Maternal Welfare Division of the Department of Public Health and carried out a series of investigations of infant mortality which supported the view, oft-expressed by contemporary reformers, that urban conditions were adverse to infant life (Scantlebury 1927b:409). Beginning in the inner city and working in co-operation with local councils she laid the foundations of the present extensive system of baby health centres. By 1927, 149 centres had been opened, 37,000 babies were being cared for each year, and the infant mortality rate had been cut from over 30 to barely 24 deaths per 1,000 live births.

Largely the work of one unusually dedicated woman, the baby health centres were arguably the most successful phase of the entire progressive movement to improve the lot of the city-bred child. Yet their accomplishments should not be exaggerated. The rate of infant death had already been on the decline for twenty years before the first health centres were founded. Other progressive reforms, such as the purification of the city's milk supplies and the promotion of elementary hygiene, may also have played a part. So may more gradual changes in housing, sanitation and obstetrical practice (Campbell, 1929-30:1-7; and compare Lewis, 1980).(13) Even when the reformers' objective was achieved, they could not take the credit themselves.

The Great Depression and the Crisis of Liberal Reform

In proportion to their efforts over the first quarter of the century, the liberal reformers could claim few clear victories. Even before the onset of the depression there were signs that their efforts were slackening and that the returns were diminishing. The number of children in kindergartens remained static and leaders of the movement complained that the children most in need of care were simply not being reached. In the late 1920s Vera Scantlebury, honorary medical officer of the Free Kindergarten Union, found disturbing levels of malnutrition in areas hard hit by unemployment and with it a marked increase in the number of 'tired, nervy, excessively emotional children' (Free Kindergarten Union, 1927-28:7-10; 1930-31:7).

If the reformers' efforts seemed unavailing it was, in part, because the forces of urbanization were working steadily against them. Between the turn of the century and the Second World War the population of Melbourne's inner core increased by about ten per cent but in the same period factory employment in the area grew by almost 350 per cent (Davison, 1979; Lack, 1980). Residential densities and problems of overcrowding reached their peak in the first quarter of the century. With the coming of the depression the old problems of overcrowding and disease were compounded by unemployment and homelessness. At the 1933 census the inner suburbs of Collingwood and Fitzroy had unemployment rates more than three times as high as some outer middle class suburbs. In one inner city kindergarten it was reported that 'the main topic of news each morning is as to whose father has a job for the day'. More than half these fathers had no work and a further quarter were only part-employed (Free Kindergarten Union, 1932-33:16). As economic prospects declined and their own environmental schemes became harder to support, the reformers were driven to a more pessimistic and, in their terms, more 'radical' social diagnosis. Having attempted, in turn, to ameliorate the urban environment of the adolescent, the child and the infant, they embraced what we must surely regard as the ultimate eugenic solutions: the forcible removal of children from the slum environment, the razing of the slums and the prevention of the 'unfit' from reproducing themselves. The first two of these measures are associated particularly with the name of F. Oswald Barnett and the third with Victor Wallace's Eugenics Society.

Barnett's campaign to abolish Melbourne's slums both culminated and largely displaced the child-saving movement of the previous twenty years. He broadened the appeal of the movement beyond a tight professional enclave to capture both the moralizing middle class and a section of the organised working class. To the eugenists' concern for the future of the city race he added a strain of old-fashioned wowserism (he first visited the slums as a Methodist campaigner against cigarette smoking) and a dash of radical economics (Barnett, 1934a:5; Russell, 1975:23; Phillips, 1981:Ch. 2). In his tract on The Unsuspected Slums, a popularised version of his 1931 M. Comm. thesis, Barnett portrayed the slum first of all as the residential expression of an unskilled, casual labour market. 'Poverty', he explained, 'is a necessary corollary to broken labour, and with poverty comes insecurity, drabness, misery and a consequent loss of morale that makes for slovenliness and vice.' A questionnaire administered to residents of South Fitzroy showed an undue proportion of 'dissolute', 'drunken', 'slovenly', 'derelict' and 'criminal' types inhabiting the most densely occupied industrial areas (Barnett, 1934a:11,33-4). Children reared in such surroundings inevitably went the way of their parents. 'Babies soak up their environment as a sponge soaks up water', he suggested, 'and a baby in a slovenly or vicious environment must, by the very laws of its being, become "slum-minded".'(Herald, 23rd May 1933, 16th January, 1934). In proof of this conclusion he cited the fact that more than 90 per cent of unfit parents had themselves been reared in similar environments. Barnett's diagnosis of the slum problem was thus an eclectic mixture of economic, biological, environmental and moral factors:

> The present social system demands unskilled casual labour. The labourer of this type ... receives poor wages. Poverty forces him into slum areas, the sordidness of these areas, plus the idle time he has, encourage evil habits ... into these hopeless surroundings little children are born, who from their birth appropriate the colour of their surroundings ... [economic necessity forces them at an early age to work] and thus the ruddy ring of roses goes on and on and on (Barnett, 1934a:25).

'How can this vicious circle be broken?' he asked. 'Only by the removal of the baby from its slum environment as soon as possible after its birth'. Barnett and his team of young Methodist laymen founded the Methodist Babies' home with the specific purpose of transplanting slum babies into stable and virtuous foster-homes.

ABOVE Children's 'Playground.' A slum street in Richmond photographed by Barnett in the 1930s.

OPPOSITE 'We are what we are because of where we are.' Slum reformers, like Oswald Barnett, saw a direct link between the slum environment and the alleged unfitness of the emerging Australian race.

Permission to reproduce these photographs from O. Barnett's Housing the Australian Nation was granted by Betty Blunden.

Rescuing babies, of course, did nothing to improve the environment into which they had been born - hence Barnett's growing concern with the abolition of the slums themselves. Only when the foetid lanes and alleys of the inner city had been cleared of cramped and decrepit dwellings, and the inhabitants moved out to pleasant garden suburbs, could the problem be said to be cured. Throughout the brilliant publicity campaign which Barnett waged during the 1930s the predicament of the city-bred child was constantly kept to the fore. The reports of the Housing Investigation and Slum Abolition Board (1937) which he largely inspired were illustrated with arresting photographs of little children playing amidst the squalor of back streets and alleys and led the reader to draw the obvious moral: that here, in the making, was a new generation of urban degenerates (Housing Investigation and Slum Abolition Board, 1937).(14)

Barnett's proposal to separate slum children from their dissolute parents may seem more heartless, though it was certainly no more radical, than the proposals of the Eugenics Society to induce the poor to stop having children altogether. The Eugenics society was never a large organization (selective breeding, after all, was unlikely to be a popular cause) but it gathered up many of the leading figures of the reform movement over the previous twenty years - Ada A'Beckett of the Free Kindergarten Union, Jane Greig of the School Health Service, Ivy Brookes of the Playgrounds Association and of course J.W. Barrett who belonged to practically all of them. It also included the cream of the city's medical, academic and judicial elite including such people as Sir Peter McCallum, Professor W.E. Agar, Sir Edmund Herring, W.S. Kent-Hughes, Justices Barry and Foster, Sir John Latham and Professor Macfarlane Burnet - who may be regarded as the last and most eminent inheritor of the eugenic tradition of reform in Melbourne (Minutes and Correspondence of the Eugenics Society of Victoria, nd and compare Macfarlane Burnet, 1979). The society's founder and leading spokeman, however, was a humble G.P., Dr. Victor Wallace. Wallace had returned to Melbourne on the eve of the depression after four years of postgraduate study in London where he had lately become acquainted with the birth control clinic established by Dr. Marie Stopes (Wallace, M.S.; see also Wallace, 1962:215-8). The main objective of the society was 'to bring about a state of affairs where persons of natural endowments shall have a higher, or at least not lower, birth rate than persons of inferior endowments; and to ensure that persons with gross defects of mind or body, known to show a tendency to be inherited, should be discouraged or prevented from having children' (Wallace M.S.:Ch. 2). During the late 1930s Wallace made several unsuccessful attempts to establish birth control clinics in Collingwood, Fitzroy and other inner suburbs. His schemes were strongly opposed by the Catholic Church (Dr. Mannix declared that 'the finest children he had seen on his walks were those in districts where it was planned to establish clinics'), by Labor politicians from some of the targeted areas, and by the Country Party government's Health

Minister (Argus, 20th and 25th April; 2nd and 9th May, 1939). Wallace himself was always careful to distinguish between mental defectives who should be compelled not to have children, and the poor, who should merely be encouraged not to reproduce themselves. But in the mind of the public and among some of his followers there was a tendency for the two groups to coalesce. 'We must stop the multiplication of these wretched stocks, these people who are mostly of low-grade mentality', declared one member of the society who went on to stress 'the need for the establishment of a birth control centre in a poor part of Melbourne (Minutes of the Eugenics Society, 23rd March, 1938, 12th May 1939). All the reformers' efforts to improve the nation's health by antenatal clinics, baby health centres and slum clearance would be for nought unless they controlled 'the fountain - head' - the reproduction of the unfit (Argus, 28th April, 1939). Except for one or two small-scale and semi-clandestine centres the birth control movement failed to achieve its purposes during the 1930s although the Eugenics Society itself limped along for another 25 years until it was finally wound up in the early 1960s (Wallace, 1962:218). Nevertheless the movement deserves to be remembered, if only as the logical culmination and final resting place of progressive reform in Melbourne.

Hitler's war completed the reorientation of reformist thinking begun by the economic shocks of the depression. The experience of prolonged unemployment and housing shortage forced a fresh awareness of the economic bases of environmental blight while the Holocaust dealt a seemingly fatal blow to the racialist assumptions of the eugenic reform. By the early 1940s a new generation of young academic economists, such as H.C. Coombs and Richard Downing, supplanted the medico-politicians as the leading ideologists of reform just as the Commonwealth government itself, from the time of the White Paper on Unemployment (1945), began to supplant the states and municipalities as the dominant partner in urban policy-making (see Davison and Lack, 1981; Rowse, 1978; Watts, 1980). Now, a further forty years on, as economic crisis once again fractures the reform consensus, and policy-makers again invoke the laws of nature to justify the ravages of the market, we may be in a better position to appreciate both the beguiling logic and the ultimate futility of the biological approach to urban reform.

NOTES

1. In this respect Stone's book may be instructively compared with A Child of Jago (Morrison, 1896).
2. In his 'Sidelights of the War on the Australian Character' (Bean, 1927-8:220-223) he claimed that the experience of war showed that 'the country bred man is, other things being equal, the better soldier'. 'The conditions in Australian cities', he nevertheless admitted, 'do not yet produce a noticeably smaller or weaker physical type than do our

country conditions; they are not responsible for the contrast which exists between a regiment of British yeomanry and one from West Ham'. The difference between the Australian soldier and his mate from the city lay rather 'in a country man having been bred to a habit of decision, whereas the mind of the city man was more dependent on the wills of those about him'.

3. Hicks (1975) shows that the fall in fertility was greatest in urban areas and among denominations with large middle-class components such as Presbyterians and Congregationalists. It is true that, in a small sample of birth data taken from South Australian birth registers, the fall in fertility in 'white collar' occupations was no greater than for 'apron' or 'blue collar' occupations but this he concedes, may be a statistical artifact or a peculiarity of the South Australian, as compared with the Victorian or New South Wales populations.
4. The proportions of the male population under 21 being educated in private schools or universities varied from 13.3 per cent in the suburbs of the inner core (Collingwood 11.8, Port Melbourne 10.9, South Melbourne 14.7) to 17.9 per cent in the south-eastern suburbs (Malvern 24.0, Kew 22.0, Brighton 21.7, Hawthorn 19.3 and Camberwell 15.1). (Commonwealth Census, 1933).
5. Barrett commonly invoked the concept of 'adolescence' (Barrett, 1916:7; 1918, Vol. 1:xiv) and defined it as the years between the ages of fourteen and eighteen (1918, Vol. 1:272-3). For a more explicit discussion, see Kent Hughes (1929:2-4).
6. See Mackinnon on the Children's Court Bill (Victorian Parliamentary Debates, 1906:3206-7) and compare the tabular material in Grabosky (1977:102) and McConville (1974:290-1).
7. Compare with the parallel division between the emphasis of the British and North American movements as described in Gillis (1974:143-7); and Kett (1977:223-4).
8. The introduction of anthropometric surveys into the state schools had been advocated since the turn of the century but it was only after 1910 that they were widely adopted (see Booth, 1911 and 1912; Sutton, 1911).
9. From 1925-6 the Free Kindergarten Union maintained a holiday home at Forest Hill to give slum children the opportunity of a few weeks' fresh air, exercise and constructive play. See Sweet (1929:21-2).
10. Report of the Third Annual Meeting of the Brisbane Creche and Kindergarten Association (1910) as quoted in Daniels and Murnane, eds (1980:126-7).
11. Enrolments in free kindergatens peaked at about 2,000 in 1927-28 and remained at about this level until 1945 (Free Kindergarten Union, Annual Reports).
12. King visited Melbourne in 1919, 1922, 1925, 1927, 1929, 1930 and 1931. An organization modelled on the Plunkett Society, the Society for the Health of Women and Children was founded

in 1920 and established several infant health centres, mainly in the northern suburbs. On the New Zealand movement, see Olssen (1981).

13. James Barrett was an early advocate of milk reform (Barrett, 1918, Vol. 2:6) and with F.W. Eggleston contrived an elaborate system of urban milk purification and delivery (Barrett and Eggleston, 1921).
14. Children at play appear frequently in photographs of slum housing from the turn of the century on. For Australian examples see Kelly, ed. (1977) and Geeves (1980).

REFERENCES

A'Beckett, A. (1935-1936) 'Development of Kindergarten Movement' in Annual Report, Free Kindergarten Union, 5 September:29.

Armstrong, W.G. (1939) 'The Beginnings of Baby Health Centres in New South Wales', Medical Journal of Australia, 29th April.

Austin, A.G. and Selleck, R. eds, (1975) The Australian Government School, 1830-1914. Melbourne, Pitman (Australia).

Australian Medical Journal (1945) Obituary of Sir James William Barrett, 14th July:58-61.

Bacchi, C. (1980) 'The nature - nurture debate in Australia 1900-1944', Historical Studies, 19, 75:199-212.

Baden-Powell, Sir Robert (1913) Boy Scouts Beyond the Seas. London, C.A. Pearson.

Barnett, F.O. (1933) 'Problem of the Children in Melbourne Slums', Herald, 23rd May.

Barnett, F.O. (1934a) The Unsuspected Slums, M.Comm.Thesis, University of Melbourne, Melbourne.

Barnett, F.O. (1934b) 'Slum Clearance as a Centenary Duty', Herald, 16th January.

Barrett, J. (1979) Falling in : Australians and 'boy conscription' 1911-1915. Sydney, Hale and Iremonger, See chapters 5 and 6.

Barrett, J.W. (1916) Imperial Alternatives - Alliance or Union. London, privately printed.

Barrett, J.W. (1918) 'Some impressions of various Universities' in Barrett, J.W., ed. The Twin Ideals : An Educated Commonwealth. Vols. 1 and 2. London, Lewis.

Barrett, J.W. (n.d.) 'Child Welfare', Typescript of speaking notes in Barrett Papers, University of Melbourne Archives.

Barrett, J.W. (n.d.) 'Education', Typescript in Barrett Papers, University of Melbourne Archives.

Barrett, J.W. (n.d.) 'National Character', Typescript in Barrett Papers, University of Melbourne Archives.

Barrett, J.W. (n.d.) 'Socialism', Typescript in Barrett Papers, University of Melbourne Archives.

Barrett, J.W. and Eggleston, F.W. (1921) The Reform of the Milk Supply, Melbourne, J.C. Stephens.

Bean, C.W. (1919) In Your Hands, Australians, London, 2nd ed. Cassel, see chapter 13.

Bean, C.W. (1927-1928) 'Sidelights of the War on the Australian Characters', Journal of the Royal Australian Historical

Society, 13:220-3.
Bean, C.W., ed. (1937) Official History of Australia in the War of 1914-1918. Sydney, Angus and Robinson 1936-42, 12 vols.
Bean, C.W. (1943) War Aims of a Plain Australian. Sydney, Angus and Robertson.
Bean Papers (n.d.) Unused Material: Old Efforts relating to 1905, Bean Papers, No 871, Australian War Memorial, Canberra.
Beddoe, J. (1857) 'On the Rapid Increase of Town Populations' Transactions of the National Association for the Promotion of Social Science, London.
Beddoe, J., (1861) 'On the Physical Degeneration of Town Populations', Transactions of the National Association for the Promotion of Social Science, London.
Booth, M. (1911) 'School Anthropometrics', Australasian Association for the Advancement of Science :689-696.
Booth, M. (1912) 'The Need of Educating Public Opinion on the Problem of the Feeble Minded', Australian Medical Gazette, 12th October:377-382.
Breckinridge, S., ed. (1912) The Child in the City. Chicago, School of Civics and Philanthropy. (Arno Reprint, 1970).
Brookes, I. (n.d.) 'The Guild of Play', People's Liberal Party (Women's Division), typescript in the possession of the Playgrounds Association of Victoria.
Burnet, F.M.(Sir) (1979) Credo and Comment : a scientist reflects. Melbourne, Melbourne University Press.
Campbell, Dame J. (1929-1930) Report on Maternal and Child Welfare in Australia. Canberra, Government Printer.
Daniels, K. and Murnane, M. eds. (1980) Uphill All the Way; a documentary history of Women in Australia. St Lucia, University of Queensland Press.
Davison, G. (1978) The Rise and Fall of Marvellous Melbourne. Melbourne, Melbourne University Press.
Davison, G. (1979) 'The Inner Suburbs - An Historical Perspective', Polis, 6, 2:39-43.
Davison, G. and Lack, J. (1981) 'Planning the New Social Order: The Melbourne University Social Survey 1941-1943', Australian and New Zealand Journal of Sociology, 17, 1:36-45.
de Beck, A.M. (1917) How Canada Answered the call in Canada in Khaki, Vol. 1. London, Canadian War Records Office.
Demos, J. and Demos, V. (1969) 'Adolescence in Historical Perspective', Journal of Marriage and the Family, 31, 4:632-638.
Dewey, J. (1916) Democracy and Education : an introduction to the philosophy of education. New York, Macmillan, reprint 1922 and 1948.
Dunstan, K. (1974) Wowsers; being an account of the poverty exhibited by certain outstanding men and women in such matters as drinking, smoking, prostitution, censorship and gambling. Melbourne, 1st edition, Cassell (Australia).
Fawcett, I.M. (1929) 'A Creative Experiment', Kindergarten Magazine, 5:September.
Free Kindergarten Union (Various Years) Annual Reports, Union

Offices, Richmond, Victoria.
Geeves, P.L. (1980) Phillip Geeves presents Cazneaux's Sydney 1904-1934 : photograph selection and biography. Sydney, David Ell Press.
Gilbert, B., (1965) 'Health and Politics : The Physical Deterioration Report of 1904', Bulletin of the History of Medicine, 39, 2:143-154.
Gillis, J.R. (1974) Youth and History : Tradition and Change in European Age Relations 1770-Present, Studies in Social Discontinuity, New York, Academic Press.
Grabosky, P. (1977) Sydney in Ferment; crime, dissent and official reaction 1788 to 1973. Canberra, Australian National University Press.
Gulick, L.H. (1907) 'Play and Democracy', Charities and Commons, 18:481-486.
Hicks, N. (1975) 'Theories of Differential Fertility and the Australian Experience', Historical Studies, 16, 65:578-80.
Hicks, N. (1978) 'This Sin and Scandal' : Australia's Population Debate 1891-1911. Canberra, Australian National University Press, especially chapter 3.
Hofstadter, R. (1944) Social Darwinism in American Thought. Philadelphia, University of Pensylvania Press.
Inglis, K.S. (1979) 'C.E.W. Bean', Australian Dictionary of Biography, 7:226-229.
Kapper, W. (1977) Biographical notes on Dr Vera Scantlebury Brown, First Director of Maternal and Child Welfare in Victoria, unpublished manuscript in Department of Social Work, University of Melbourne.
Kelly, M. (1977) A Certain Sydney. Sydney, Doak.
Kent-Hughes, W. (1929) 'Adolescence', Australian Educational Quarterly, 5:2-4.
Kett, J.E. (1977) Rites of Passage : Adolescence in America, 1770 to the Present. New York, Basic Books.
Kindergarten Training College (1926-27) Prospectus. Institute of Early Childhood Development Papers, Melbourne University Archives.
Kinglake, E. (1892) The Australian at Home. London, Leadenhall Press.
Lack, J. (1980) Residence, workplace, community. Local history in Metropolitan Melbourne, Historical Studies, 19, 74:16-46.
Lawson, H. (1892) 'The Captain of the Push' in Roderick, C. ed. Henry Lawson, Collected Verse, Volume 1 1885-1900. Sydney, Angus and Robertson, 1972, 2 vols.
Lewis, M. (1980) Milk, Mothers and Infant Welfare, in Roe, J. ed. Twentieth Century Sydney : Studies in Urban and Social History. Sydney, Hale and Iremonger in association with the Sydney History Group.
McConville, C. (1974) Outcast Melbourne : Social Deviance in the City, 1880-1914, Unpublished M.A. Thesis, University of Melbourne.
McConville, C. (1981) 'Outcast Children in Marvellous Melbourne', in Featherstone, G. ed. The Colonial Child. Melbourne,

Royal Victorian Historical Society.

McNab, K. and Ward, R. (1962) 'The Nature and Nurture of the First Generation of Native-Born Australians', Historical Studies, 10, 39:289-308.

Morrison, A. (1896) A child of the Jago. London, Penguin Books reprint 1946.

Murray-Smith, S. (1979) 'Sir James William Barrett', Australian Dictionary of Biography, 7:186-189.

Norris, A. (1978) Champions of the Impossible : A History of the National Council of Women 1902-1977. Melbourne, Hawthorn Press.

Olssen, E. (1981) 'Truby King and the Plunkett Society: An Analysis of a Prescriptive Ideology', New Zealand Journal of History, 15, 1:3-23.

Phillips, R.D. (1981) The Origins of the Housing Commission of Victoria, Unpublished M.A. Thesis, University of Melbourne, Melbourne.

Playgrounds Association of Victoria (n.d.) Brief History of the Association of Victoria to 1944; typescript in the possession of the Association, made available by Anne Webster.

Platt, A. (1969) The Child Savers : The Invention of Delinquency. Chicago, University of Chicago Press.

Reeder, D.A. (1977) 'Predicaments of City Children : late Victorian and Edwardian Perspectives on Education and Urban Studies', in Reeder, D. ed. Urban Education in the Nineteenth Century, Proceedings of the 1976 Annual Conference of the History of Education Society of Great Britain, London, Taylor and Francis.

Reports of the Department of Neglected Children (various years) in Victoria Parliamentary Papers. 1905, 3, 33; 1909, 2, 17; 1911, 2, 20; 1912, 3, 25.

Report of the Minister for Public Instruction (1910 and 1911) Reports on Medical Inspection and Kindred Subjects, Victoria, Parliamentary Papers, 1910, 2, 44; 1912, 2, 12.

Report of the Minister of Public Instruction (1912) Victoria, Parliamentary Papers, 2, 12.

Roth, R.E. (1914) 'School Posture and Spinal Deformities', Transactions of the Australian Medical Congress, 774-777.

Rowse, T. (1978) Australian Liberalism and National Character, Melbourne, Kibble Books.

Royal Commission on the Housing of the People of the Metropolis (1917) Evidence of Dr Harvey Sutton, Victoria, Parliamentary Papers, 2, 29.

Russell, E.W. (1972) The Slum Abolition Movement in Victoria 1933-37. Melbourne, Hornet Press.

Sandercock, L. (1975) Cities for Sale, Melbourne, Melbourne University Press, especially chapter 1.

Scantlebury, V. (1927a) 'Experiences Abroad with Special Reference to Infant Welfare', Medical Journal of Australia, 8th January, 35-37.

Scantlebury, V. (1927b) 'Address on Some Aspects of Infant Welfare Work in Victoria', Health Bulletin, 13:409.
Scantlebury, V. (1928) 'Infant Welfare Work', Medical Journal of Australia, 25th February: 251-252.
Scantlebury, V. and Main, H. (1926) Report to the Minister of Public Health on the Welfare of Women and Children, Melbourne, Victorian Baby Health Centres Association. Copy in Brownless Medical Library, University of Melbourne.
Searle, G.R. (1971) The Quest for National Efficiency: a study in British Politics and Political Thought 1899-1914. Berkeley, University of California Press.
Searle, G.R. (1976) Eugenics and Politics in Britain 1900-1914. Leyden, Noordhoff International.
Select Committee on the Housing of the People of the Metropolis (1913-1914) Victoria, Parliamentary Papers, 1, 2.
Spearritt, P. (1974) 'Middle Class Reformers and the Labour Response', Labour History, 26:65-81.
Spearritt, P. (1974) The kindergarten Movement : tradition and change, in Edgar, D.E. eds. Social Change in Australia : Readings in Sociology Melbourne, Longman Cheshire.
Springhall, J. (1977) Youth, Empire and Society - British Youth Movements 1883-1940. London, Croom Helm.
Stedman Jones, G. (1971) Outcast London : A study in the relationship between classes in Victorian society. Oxford, Clarendon Press.
Sutton, H. (1911) 'The Feeble-Minded - their Classification and Importance'. Transactions of Australian Medical Congress, 894-905.
Sweet, Dr. (1929) 'Address to holiday Home Annual Meeting' Kindergarten Magazine, 5:September.
Tanner, T.W. (1980) Compulsory Citizen Soldiers. Sydney, Alternative Publishing Cooperative.
Twopeny, R. (1883) Town Life in Australia. Sydney, Sydney University Press, 1973, Facsimile reprint of 1883 edition.
Veiller, L. (1907) 'Social Value of Playgrounds in Crowded Districts' Charities and Commons, 18:509.
Waite, F. (1921) The New Zealanders at Gallipoli. Auckland, Whitcombe and Tombs Ltd.
Wallace, V.H. (1962) 'The Eugenics Society of Victoria (1936-1961)', The Eugenics Review, 53, 4:215-218.
Watts, R. (1980) 'The origins of the Australian Welfare State', Historical Studies, 19, 75:175-198.
Wiebe, R. (1967) The Search for Order, 1877-1920. London, Macmillan.

8 The development of urban planning in Australia 1888–1948: a bibliography and review

Robert Freestone

The study of the history of urban planning has been justified in a number of ways. In 1943, Feiss argued that 'in a field changing as rapidly as city planning, textbooks become obsolete overnight, but the constantly enlarging material dealing with historical background remains an important basis for the reasoned steps any planner must take' (Feiss, 1943). A similar utilitarian logic is adopted by Sutcliffe (1981a:7). There is also the academic stance: the past not only informs the present but enriches our understanding of the 'spirit and purpose of planning' (Cherry, 1969). Hancock sees the historical analysis of planning mitigating professional ignorance, offering a more secure basis for decision-making 'than that provided by knowledge of the present alone', fostering a critical reappraisal of basic concepts, and explicating 'the whole process of planned innovation and societal change' (Hancock, 1972). Eschewing elaborate justification Reps puts his case simply and bluntly: 'there is really no need to be defensive about planning history. It obviously has merit' (Reps, 1981:13). One index is the 1400 titles assembled in a recent bibliography and adjudged 'by any standards ... an impressive body of materials' (Sutcliffe, 1981a:8).

The case for an Australian planning history needs no peculiarly nationalistic foundation, although the remarkable urbanization and suburbanization evident since the nineteenth century and the view that the history of Australian cities is that of 'a multiplicity of plans' (Harrison, 1977:6) adds a special interest. Although planning history is yet to formally congeal as an identifiable field of interest in Australia, a respectable body of relevant literature does exist and the future looks even more promising. The growing international interest in planning history will undoubtedly be one catalyst for local research, as will the continuing development of Australian urban history (Davison, 1979). Planning as a profession also seems headed for a period of critical reassessment and thus a rediscovery of the pertinence of the past, a mood particularly evident at the 1980 Planning Institute Congress (Royal Australian Planning Institute Journal, 1980).

Against this background, the present paper offers a first

guide to the literature of Australian planning history. There are three main sections: a brief, selective review of recent work; some thoughts on directions and desiderata for future research; and thirdly, and forming the basis for these two sections, presentation of a short bibliography.

The temporal distribution of Australian studies shows a similar pattern to the international literature. The pre-1914 and post-1945 periods attract most attention and the 1920s and 1930s are more of a 'grey area' (Sutcliffe, 1981a:9). Colonial town planning is an established, well-defined research area in Australia.(1) The last thirty years are well covered, at least in outline, although there is the difficulty of identifying relevant work from a literature which is almost inherently 'historical'.(2) The work reviewed in this paper pertains mainly to the development of physical planning over six decades from the late 1880s to the late 1940s. This period saw a remarkable evolution: the emergence of urban planning in its modern form; spirited propagandist activity around the First World War and subsequent passage of pioneering legislation; a less heady phase during the prosperous 1920s when business-like planning thought predominated; and from the early 1930s a more critical and professional approach culminating in a high profile for urban (and regional) planning in the reconstructionist fervour of the 1940s. The chronology of events in Table 1 summarizes this development, symbolically launched by the fin-de-siecle discovery of the model suburb idea and capped by the presentation of the country's first statutory metropolitan plan. It is an optimistic start and finish to the formative years of Australian planning, but tempered in the intervening period - and since - by failures.

The writing of Australian planning history: a review

Urban planning history in Australia exists as a diverse but not voluminous literature, the product of varied objectives. There is a sprinkling of outstanding, stimulating works but there are few conscious contributions to the field. Mercifully no 'urban planning historians' have appeared. The nature of the literature covering the formative years is to some extent dependent on the precise specification of that literature. Some general features characterize the appended bibliography.

TABLE 1 Urban Planning In Australia 1888-1948: A Selected Chronology

1888	Harcourt model suburb, Sydney.
1889	J. Keily's 'Study on Unity of Design in Planning New Towns and Suburbs' read before Victorian Institute of Surveyors Kensington model suburb, Sydney.
1890	J. Sulman's 'The Laying Out of Towns' read before A.A.A.S. meeting, Melbourne. Melbourne and Metropolitan Board of Works.
1901	Congress of Engineers, Architects, and Surveyors on the laying out and building of the Federal Capital, Melbourne.

G.H. Knibbs' 'The Theory of City Design' read before the Royal Society of New South Wales.

1907 J. Sulman's The Improvement of Sydney.

1909 Report of the Royal Commission for the Improvement of Sydney and Its Suburbs.

1910 J. Sulman's 'The Federal Capital of Australia' read before the Town Planning Conference of the Royal Institute of British Architects, London.

1912 W.B. Griffin wins Federal Capital design competition Dacey Garden Suburb developed by New South Wales Housing Board.

1913 New South Wales Town Planning Association. Report for the New South Wales Government of the Commission of Inquiry (R.F. Irvine) into the Question of the Housing of Workmen in Europe and Australia.

1914 G. Taylor's Town Planning for Australia. W.E. Bold's Report on Tour Round the World for Perth City Council. Town Planning Association of Queensland. Town Planning section at Australian meeting of the British Association for the Advancement of Science. Australian Town Planning Tour by C.C. Reade and W.R. Davidge for the Garden Cities and Town Planning Association.

1915 South Australian Town Planning and Improvement Association. Southern Tasmanian Town Planning Association. Town Planning and Housing magazine. Town Planning: report to the Victorian Minister of Public Works by J.C. Morrell.

1916 Vernon Memorial Lectures in Town Planning at Sydney University. Town Planning Association of Western Australia. C.C. Reade appointed Town Planning Adviser to the South Australian Government.

1917 First Australian Town Planning and Housing Conference and Exhibition, Adelaide.

1918 Second Australian Town Planning Conference and Exhibition, Brisbane. G. Taylor's Town Planning with Common-sense. New South Wales Town Planning Advisory Board.

1919 First Victorian Town Planning Conference and Exhibition, Ballarat. New South Wales Local Government Act (provision for residential districts). South Australian Garden Suburb Act. C.C. Reade's Planning and Development of Towns and Cities in South Australia.

1920 South Australian Town Planning and Development Act.

1921 J. Sulman's An Introduction to the Study of Town Planning in Australia.

1922 Sydney Regional Plan Conference.

1923 Melbourne Metropolitan Town Planning Commission. Queensland Local Authorities Amendments Act.

1925 W.J. Earle appointed Town Planner for Greater Brisbane Council.

1928 Western Australian Town Planning and Development Act. Metropolitan Town Planning Commission, Perth.

1929 Plan of General Development, Melbourne: report of the Metropolitan Town Planning Commission.

1930 Report of the Metropolitan Town Planning Commission, Perth.
1931 Town Planning Institute of Western Australia.
1934 Town and Country Planning Institute of New South Wales. Queensland City of Mackay and Other Town Planning Schemes Approval Act
1938 Victorian Housing Commission.
1941 New South Wales Housing Commission.
1944 Final Report of the Commonwealth Housing Commission. F.O. Barnett, W.O. Burt and F. Heath's We must go on: a study of planned reconstruction and housing. Tasmanian Town and Country Planning Act. Victorian Town and Country Planning Act. Planning Institute of Australia, Melbourne.
1945 W. Bunning's Homes in the Sun: the past, present and future of Australian housing. New South Wales Local Government (Town and Country Planning) Amendment Act.
1947 Town Planning Institute of South Australia.
1948 British Council Town Planning Exhibition. Visit of Sir Patrick Abercrombie. D. Winston appointed to Chair of Town and Country Planning at Sydney University. Report on the Planning Scheme for the County of Cumberland, New South Wales.

Three different types of author are represented: pre-professionals (mainly student architects, planners and historians), practising planners, and the usual cast of academics drawn primarily from architecture, geography, planning, political science, and urban studies. Planning history is a multi-disciplinary endeavour but there is some imbalance. Historians have generally kept a low profile, yet have produced some of the more lasting contributions such as those on city improvement in the nineteenth century by Laverty (1968), Mayne (1980a) and Roberts (1978a). Local histories occasionally touch on town planning and their typically factual content can prove especially valuable when related to more general developments. Bate's history of Brighton, Victoria has both elements and indicates the importance and limitations of planning at the local level from the colonial era (Bate, 1962: 30-39, 238-242 & 380-385). Surprisingly, there are few distinguished contributions from planners. Their students are only a little more active. Perhaps planning history, especially Australian planning history, is only an obligatory preamble to the 'real' substance of planning education?

At the First International Conference on the History of Urban and Regional Planning in 1977 three broad interpretative frameworks were evident: the marxist, the functionalist, and the dominant liberal-progressive (Sutcliffe, 1981b:3). Marxist interpretations are virtually absent in Australia planning history although snippets appear as part of a more general story in Connell and Irving (1980:191, 205). Post-war planning is now attracting a radical critique (Mullins, 1979, Orchard, 1979), but for the period under review the most provocative account is by Sandercock (1975) whose mild political economy is still radical enough to seem marxist to some critics (e.g., Grubb, 1976a:61). Most Australian work falls into either the functionalist or, and more commonly when any real analysis is involved, the liberal-progressivist mode of argument. To a certain extent these are secondary labels for a dichotomy in the literature between the 'insider-planner' versus the 'outsider-social scientist' contribution. Neither questions the fundamental 'goodness' of planning but the former adopts an 'objective' factual stance concentrating on a chronology of events while the latter is more likely to be value-critical, setting such events more carefully in a broader social, economic and political context.(3)

Empiricism is a major mark of the literature, although the number of contributions which actually uncover fresh data is surprisingly limited. There is a derivative quality about many works (especially the entries in section A2 of the bibliography). Reliance on secondary sources need not produce stale results, as evidenced by the entertaining accounts of planning developments woven into populist texts by Spearritt (1978) on Sydney (4) and Stannage (1979) on Perth.

A diligent reader could piece together a detailed but by no means exhaustive overview of planning in Australia from the references supplied here. But there is nothing remotely comparable to definitive overseas studies by Ashworth (1954), Cherry (1974), and Scott (1969). Even a summary overview is hard to find though O'Brien's entry on town planning for the 1958 Australian Encyclopaedia is still a useful and accurate survey of pre-1950 developments.(5) Engaging essays by Freeland (1972), Ledgar (1976) and Winston (1976), and chapters by Powell (1976) and Bolton (1981) in their respective books on historical aspects of environmental policy in Australia offer the most readable accounts of the beginnings of town planning. Sandercock's Cities for Sale is often read as a critique of contemporary planning but the first half of the book is concerned with planning before the Second World War.(6) Citations in other work published since 1976 confirm it as the standard reference on the early years. Although an ambitious and quite seminal treatment, as planning history it is somewhat flawed by a broadbrush approach (Adelaide, Melbourne and Sydney since 1900 in a little over 200 pages) and the basic *modus operandi* which shows the author's central interests to be ultimately elsewhere.(7) Inevitably, many events must be treated superficially, and later writers have disagreed with particular interpretations.(8)

The full record of planning events, discussion of overseas influences on local planning thought, planning ideology, and urban reform prior to 1900 are some of the important questions not pursued far enough in Cities for Sale.(9) The treatment of individual planners in chapter one ('Emergence of Town Planning Movement 1900-1920') cannot be accepted uncritically. Two dominant strands in the early planning movement are identified: the 'aesthetic' and the 'social' represented by Sir John Sulman and J.D. Fitzgerald respectively. This dichotomy is also restated by Powell (1976:133) but it ignores their shared commitment to growth, (Sulman, 1927:7-8; Fitzgerald, 1918:2-3) private property (10), the city beautiful (Sulman, 1921 and Fitzgerald, 1914a), and eugenics (Sulman in evidence to the Royal Commission on the Housing Conditions of the People in the Metropolis 1917, 299 and Fitzgerald, 1914b:10). Professor R.F. Irvine, the Commissioner for the New South Wales Government (Table 1) emerges not only on the 'social' side but the one planning advocate supporting 'fundamental structural changes'. This tends to overstate and misrepresent the uniqueness of Irvine's position. A more exhaustive reading of his work would reveal an unremarkable support for suburbanization, technocratic planning and environmental determinism (Irvine, 1914 and 1915), a passionate aestheticism (Irvine, 1916), and an uncompromising honesty which as much as anything else brought him into conflict with others - even on the 'social' side of the movement (Surveyor, 1914:22). Despite these selective criticisms Cities for Sale remains an important book full of interesting leads and hypotheses and its author deserves most of the credit for introducing 'some new figures into the mainstream of Australian history, the town planners' (Roe, 1976).

The history of planning in individual states is dominated by that of their capital cities, and so Sydney is the focus of the general histories of New South Wales by Larcombe (1978), Maiden (1960) and Migliorino (1976). The historical coverage for paticular capitals varies. Much literature on Canberra qualifies as a kind of planning history and the bibliography by Johnson (1980) assembles pertinent references to 1976. Nevertheless a comprehensive and critical history of physical planning and other matters such as the influence of the Canberra experience remain to be tackled. The story of urban planning in Brisbane since the nineteenth century can be traced through several substantial works, starting with Laverty (1954 and 1968), and on to Cox (1968) and then Fletcher (1977) for post-war developments. The Melbourne scene from the 1900s to the 1940s is covered comprehensively by Grubb (1976a) who charts the progress of planning through five phases: the diffusion of ideas 1910-1915, the consolidation of the movement 1915-1920, the tenure of the Metropolitan Town Planning Commission, the partnership of the slum abolition and planning movements 1930-1943, and a final phase which saw the enactment of the Town and Country Planning Act in 1944. Grubb unambiguously accepts the liberal-progressivist perspective; planning is seen as 'a logical extension of (the) general movement of government assuming wider community responsibilities' (Grubb, 1976a:3; see also Grubb, 1976b:5). No other Australian contribution matches the coverage and detail of this work although it is unfortunately not without errors.(11) A more serious shortcoming is the author's empiricist bent which begs the crucial question of planning by whom for whom and results in an uncritical reporting of many events and some glaring contradictions. There are admiring references to 'publicly minded citizens' but no consideration of 'the public interest'. The 'most enduring contribution' of the Town Planning Commission was 'in an educational capacity' yet the deferring of planning legislation for another thirteen years is partly attributed to 'a lack of understanding indeed apathy on the part of governments'. And at the climax of the story the scathing comments of the leading reformer, W.O. Burt are reproduced without comment. To him ,the Victorian legislation of 1944 was a 'miserable measure', 'ill-considered', and the 'still-born creation of English legislation which ... has been an admitted failure' (Burt, 1945).

Analyses of individual planners, planning theory and planned environments are less constrained by the legislative record unlike general histories. There are already some interesting contributions in this broad area (see mainly section A3 of the bibliography), and one is left with the impression of considerable scope for future work. Clark's (1969a) pioneering study of city beautiful thought in Perth could be usefully followed up in other states. The garden city idea was also influential in the west (Clark, 1969b; Webb, 1968 and 1979) and right across the country (Freestone 1981). The so-called 'garden suburb' was the most popular Australian interpretation before the Second World War. Johnson (1976) and Crow (1978) look at Sydney's 'first' garden suburb at Haberfield, although such estates were anticipated by model suburbs in the nineteenth century (Peek, 1976; Roberts, 1970). Price's (1969) thesis on the first state showpiece at Daceyville is the best of several unpublished studies. A meticulously detailed sociological investigation of the development of community life in Charles Reade's model suburb of Colonel Light Gardens in Adelaide is provided by Henry (1955). One of the more intriguing schemes was the plan for an 'Australian Bournville' near Hobart. Shelton <u>et.al</u>. (1973) has discussed the settlement as a physical plan, while Barton (1981) concentrates on Claremont as an experiment in industrial welfare. Probably the best case study to make explicit the links between town planning, welfare capitalism and social control is by Burrows (1978) on Lutana, the E.Z. Company's garden village, also on the Derwent River. The 'garden suburb' went into production in the 1920s as a real estate concept and Docherty (1977) examines this process in Newcastle.(12) There were few distinguished projects in this era. The master plan for the City of Perth Endowment Lands by Hope and Kelm (Selwood, 1979) and Walter Burley Griffin's 'land planning' experiment at Castlecrag (Bryant, 1967; Harvey, 1964; Harrison, 1973; Johnson, 1971) were two notable exceptions.

Griffin has rightly attracted a considerable literature, although only the thesis by Harrison (1973) on this 'prophet without honour' concentrates on his city and suburb plans (c.f., Birrell, 1964; Johnson, 1977). Yandell's (1980) biography of Dr. J.J.C. Bradfield of Sydney Harbour Bridge fame is the most important study of a figure allied to the early planning movement. J.M. Freeland at the University of New South Wales supervised a number of biographical studies which refer to the planning activities of their subjects - mainly early architects. The resultant theses on individuals such as Sulman (Humphrey, 1968), Reade (Brown, 1978), Sydney 'City Beautiful' advocate J.B. Barlow (Somers, 1970), New South Wales Government Architect W.L. Vernon (Jones, 1977), and George and Florence Taylor (Murray, 1976; van Rooijen, 1976) are not definitive works but they all contain much valuable information. Murray's study of Florence Taylor, one half of 'possibly the most amazing couple in Australia's history' (Freeland, 1971:78), runs to three volumes and even reproduces another unfinished biography by Kerwin Maegraith. The most brilliantly packaged biographical study remains that by George (1971) on Henry Halloran - zoning advocate, surveyor, and speculator. Like Robertson's (1970) study of Perth Town Clerk W.E. Bold, Petrow's (1978) thesis on Fitzgerald encompasses more than the subject's planning activities and also profitably borrows from American perspectives on the Progressive era. Other planners and planning advocates have been much more summarily dealt with (13), if at all.

Several general conclusions emerge from my admitedly selective reading of the literature. Large gaps in coverage remain, there are few definitive studies, and the literature is probably most impressive in quantitative terms. More importantly, at least three major themes have already surfaced. First, the importance of British precedent in early Australian developments is stressed by most writers. Second, even the shift from idealism to pragmatism in planning thought by the 1920s could not bridge the persistent gap between theory and practice. Third, and underlining the importance of the period under review, many of the basic tasks for planning in Australia were cystallized early. Gibbons (1972 and 1980) sees the 1908-1909 New South Wales Royal Commission on Sydney (Table 1) as a 'watershed' in the planning of the city, echoing Winston's (1957:26) remark that 'nearly every subsequent improvement to the central area of Sydney ... was suggested and clearly explained in this report'. The same theme is reiterated in Webb's important study of Perth where 'there remains to this day hardly a detail about planning ... which cannot in some way be traced back to a genesis in the debates and reports of the period 1899-1914' (Webb, 1979:381). The strucural constraints on planning in a capitalist society were also evident early. The 1916 Town Planning and Housing bill in South Australia - the first but unsuccessful planning legislation introduced into an Australian parliament - 'within a week ... had raised some, and within six months most, of the dilemmas that have (since) dogged the progress of planning' (Sandercock, 1975:37).

The past in the future: some general thoughts

The long term prospects for planning history in Australia are assured. Perhaps even the bibliography of this paper will date quickly in the short term. Didactic recommendations for future research into the formative years of Australian planning are unnecessary but some broad areas worthy of exploration do stand out. For a start, the first two general themes which are becoming accepted demand closer scrutiny.

As for the pervasive influence of British ideas, the mechanics of the diffusion process and the uniqueness or otherwise of the 'export' campaign to Australia should be investigated more fully.(14) Indeed, the international frame of reference must be widened. Grubb (1976a) highlights American influences on the style of Melbourne planning in the 1920s as well as the importance of the 1926 New Zealand Town Planning Act in general propaganda and specific legislative proposals. Continental initiatives were followed with interest. The distinguished German planner Werner Hegemann visited Australia before the First World War (Fitzgerald in evidence to the Royal Commission on the Housing Conditions of the People in the Metropolis, 1917:304), and even Camillo Sitte was invited to prepare city plans for Adelaide, Melbourne and Sydney (Collins and Collins, 1965:12). The Canadian connection has been totally overlooked. Reade, who corresponded frequently with Thomas Adams (Artibise and Stelter, 1979), emphasized that 'Canadian legislation and conditions are nearer to our own than are the conditions in any other part of the world' (Report of the Select Committee on the Town Planning and Development Bill, 1920). The fusion of city beautiful and garden city influences in Griffin's Canberra plan (Peisch, 1964) suggests an eclectic flavour to early Australian planning thought. When 'planning' is substituted for 'architecture' the central theme of the most recent history of modern architecture in Australia could therefore serve as one guide to future research: 'the migration of ideas on the art of (planning) and their adoption in the Australian environment' (Johnson, 1980:205; see also Ledger, 1976 and Winston, 1976). The distinctiveness of Australian planning - an internationally informed response to relatively mild urban problems in an economically and culturally dependant society - might then be appreciated.(15)

A second broad area in which to set future research is the debate on the 'failure' of Australian planning. For those who accept the proposition, the case probably starts with the demise of Governor Phillip's ambitious Sydney plan of 1788 (Historical Records of Australia, 1914), strengthens via a chronology of lost opportunities and unrealized promises into the twentieth century (Sandercock, 1974 and 1975) and is probably sealed by A.J. Powell's (1980:5) recent epitaph that 'after more than three decades of statutory planning the public benefits seem to be very limited indeed'. Reflecting the insider-outsider dichotomy in the literature, some favour a political explanation for the lack of success, while others look to technical and organizational failings. Harrison (1978:142) integrates the two in explaining the modest achievements of early statutory planning: 'there was a general reluctance to interfere with the rights and consequently the values attaching to property ownership and the procedures required to prepare and administer a statutory plan are necessarily ponderous'. Historical analysis of this issue must also consider Australia as the 'lucky country'. The pursuit and attainment of home ownership was basic to the country's high standard of living even before the turn of the century and 'this atomistic growth strangled planning at birth without any outcry' according to Fry.(16) A structuralist interpretation offers yet another explanatory approach. Writing on the North American experience but with direct relevance to Australia, Scott and Roweis claim that the deficiencies of urban planning 'emerge not so much as the result of an analytical failure, but as the inevitable consequence of a refractory (though mutable) social situation that at once clamours for, and yet rebuffs, collective urban intervention' (Scott and Roweis, 1977; see also Scott, 1980).

The merging of various strands of urban reform into the cohesive approach to urban management formally termed 'town planning' by about 1909 is another important area for future research. Haratsis (1979) criticizes written history which treats planning as essentially a twentieth century phenomenon. He sees modern town planning emerging in Melbourne during the 1870s, in particular with the 1873 Report of the Lowlands Commission on the location of noxious industry. The basic continuity of the quest for effective land use control goes back much further of course (see note 1) but a really critical phase seems to be the 1880s which saw marked metropolitanization, suburban land booms, severe strains on transportation and sanitation systems, and colonial economies moving to the crest of a slump. Roberts' (1978a) seminal study of Sydney shows how the scale, complexity and interdependence of urban problems came to be appreciated during this 'age of improvement'. Interrelationships between physical planning and other social reforms - especially direct responses to urbanization such as metropolitan government, anti-slum campaigns, decentralization and the national parks movement - need to be unravelled.

These suggestions certainly do not exhaust the agenda. Solid biographical research is badly needed, for example. John Sulman is an obvious candidate. He gave himself (17), and has been accorded by others (18), the status of lone ranger. But significantly 'The Laying Out of Towns' (1890) not only synchronized with overseas developments but referred to model suburbs on the drawing board for Kensington in Sydney and Hopetoun on Melbourne's northern outskirts (Sulman, 1890). Sulman's call for more enlightened civic design and tighter land use control was also anticipated by contemporaries.(19) The life and work of the early planners should not be uncritically accorded heroic status. By viewing their subjects in the context of the ideology of reform, Petrow's (1978) thesis on Fitzgerald and to a lesser extent Hyslop's (1980) treatment of Sir James Barrett in Melbourne show how this can be done. The minor characters are not without interest, but the history of planning is more than the sum of individual efforts. The metamorphosis of voluntary reform movement to profession evident as early as the Brisbane Town Planning Conference (Table 1) is the note on which

Powell's (1976:166) discussion so intriguingly ends. This transformation and the subsequent story of the planning profession in Australia remains to be told in full.(20)

A richer appreciation of the evolution of Australian planning requires careful historical research, preferably based on primary materials when available. There may be some finite limit on available sources but even relatively accessible data are not being used to advantage. The history of federal urban policy by Lloyd and Troy (1978) for example excludes at least two important calls for national intervention before 1920. In 1919, Reade lobbied for a national 'Civic Welfare Bureau' similar to the Town Planning and Housing branch of the Canadian Commission of Conservation. And predating the Commonwealth Housing Commission by more than two decades, the Inter-State Commission declared that the problems of urban Australia 'cannot be solved without prompt action, including some financial assistance on the part of the Commonwealth Government. The mischief is nation-wide and needs a national remedy'.(21) More diligent empirical research alone will not lead to maximum progress. An infusion of theoretical perspectives will also help focus attention beyond surface details and onto more fundamental questions and interrelationships in a wider social context. Two potentially rewarding approaches are already apparent. One is the further development of an explicitly 'Progressive' interpretation of early twentieth century urban reform. Certainly the seemingly incongruous membership of Australia's first town planning association with its 'colour range from the purest of blues to a most violent crimson' (Powell, 1976:161) and other differences between planning advocates may be at least partly resolved by their shared Progressivism/Liberalism with its emphasis on efficiency, rationality, cooperation, and national welfare.(22) Greater insight into the ideological ramifications of Australian planning might secondly be facilitated by examining the axiom of radical theory that the historical function of planning has been 'to benefit local, regional and national business rather than the broader public' (Claval, Forester and Goldsmith, 1981:vii).

CONCLUSION

In An Introduction to the Study of Town Planning in Australia (1921) John Sulman surprisingly dealt with formal planning history in one brief introductory chapter. Brown and Sherrard in their sequel Town and Country Planning (1951) did much the same thing. F.J. Osborn labelled their treatment 'rather perfunctory' and asked exasperatingly: 'when *will* planners learn that history is a study in itself?' (Osborn, 1952). He would have found little joy in a revised version of the book published nearly two decades later (Brown, Sherrard, with Shaw, 1969). This reviewer can now be more optimistic. There is clearly a growing critical interest in history. The generic term 'urban planning history' implying a coherent and substantive body of work may be a little premature. But the achievements so far still augur well for the future of the past.

NOTES

1. See for example: Atkinson (1981), Bate (1962), Burchett (1979), Davey (1969), Glasscock (1967), Hugo-Brunt (1975), Jeans (1965, 1967, 1981), Laverty (1971), Matthews (1968), Safran (1977), Steele (1975), Van Zyl (1963), Walkley (1954), Watts (1976), Wheeler (1966, 1967), Williams (1966a,b, 1974).
2. But see for example section A2 of the bibliography and Alexander (1981), Fagence (1979), Gibbons (1976), Harris and Dixon (1978), Harrison (1971, 1974), Lloyd and Troy (1981), Logan (1981), Powell (1969), Troy (1978).
3. Compare for example Walkley (1952) and Williams (1974) on Adelaide, Gawler (1963) and Bowman (1976) on Melbourne, Poole (1979) and Webb (1979) on Perth, and Colman (1968) and Spearritt (1978) on Sydney.
4. The transformation from dissertation (Spearritt, 1976) to book (1978) was accomplished with few changes. The one significant section of the dissertation dealing with planning and omitted from the book covers a controversy over building height in the C.B.D. involving John Sulman and also the lobbying work of the Sydney Regional Plan Convention (or Conference) in the 'grey' 1920s (61-63).
5. Although superseded by Harrison (1977b).
6. The book preserves the basic structure of the dissertation. Chapter 4 of the dissertation ('Sydney: House, land and car: consolidation of the "first suburban nation"') undergoes a major revision which substitutes an explicitly political coverage in line with the other chapters for nonetheless useful material on the development of planning ideas. Lost in the shuffle once again is reference to the Sydney Regional Plan Convention.
7. 'This (study) does not assume as given the pluralist-democratic model. Rather it inquires in each of the situations and issues analyzed as to where ultimate power lies, and what this suggests about the nature of Australian society' (Sandercock, 1974:viii). See also: Sandercock (1978).
8. For example: Gibbons (1980:132) argues that dismissal of the recommendations of the 1908-1909 New South Wales Royal Commission (Table 1) as a confused pot-pourri unlikely to have resulted in concerted action 'lacks logical or empirical foundation'; Roberts (1978a:373) criticizes the implicit acceptance of the simplistic parties of initiative (Labour) and resistence (non Labour) dichotomy common in Australian political history; and Grubb (1976a:61-63) argues that attributing the demise of the Melbourne Metropolitan Town Planning Commission 1930 to the instability and Country party hostility of Victorian politics overlooks the limited mandate of the Commission and the short-lived municipal

support to be expected.

9. On each of these points: What were the 'numerous boards and commissions' through which John Sulman had such limited success? Was Australian planning thought only 'heavily derivative of British thinking and values'? Could not the environmentalist assumptions of the planning advocates be seen more charitably given that 'basic physical necessities such as decent housing and a pleasant environment may not be a sufficient condition for social progress, but they are quite likely to be a necessary condition'? (Broadbent, 1977:149). And is Sulman's 1890 paper 'The Laying Out of Towns' read before the Australasian Association for the Advancement of Science the only nineteeth century bellwether worth mentioning?

10. J. Sulman gave evidence to the Royal Commission on the Housing Conditions of the People in the Metropolis, (1917:301). J.D. Fitzgerald gave evidence to the Parliamentary Standing Committee on Public Works, Inquiry into the proposed scheme for housing Commonwealth workmen at Lithgow, (1917-18:85). Fitgerald 'came out' at this inquiry to state that he 'was never a leasehold man myself, although I accepted the principle when I was a member of the Labor Party for the bigger things I could get '. To complicate matters, Sulman at the same inquiry conceded that public ownership might be advantageous.

11. For example: the Illustrated House and Property Catalogue (1896-1897) was a real estate directory and not a professional journal (p. 10); Thomas Mawson never made it to Australia in 1913 (p. 11 and 13); J.W. Barrett's paper on town planning was presented to the Imperial Health Conference in May 1914 (p. 11); the Australian Town Planning Tour began in August 1914 and Charles Reade continued the lectures into 1915 (p. 14); the South Australian Town Planning and Housing bill was introduced in 1916 (15); subscriptions to English journals would not necessarily explain why the Royal Victorian Institute of Architects published articles by Sulman in their journel (p. 17); Building magazine was founded in 1907 and not 1911 (p. 18), etc.

12. This study considers other aspects of planning in Newcastle between the wars, most notably A.G.C. Tewksbury's 1926 remodelling scheme (180-184).

13. Other biographical notes: include Harrison (1977), McFarlane (1966), Oldham and Oldham (1980), Spearritt (1977), Stannage (1977).

14. The pre-1939 channels for diffusion of information discussed in general studies such as Grubb (1976a), Powell (1976) and Sandercock (1975) are local and international journals, official lecture tours, and overseas 'fact-finding' missions. Personal connections, schemes such as the travelling scholar programme of the Board of Architects of New South Wales (see D.L. Johnson, 1980) and war service

were among other doors to the world. On the importance of military service abroad see: Rudduck and Grounds (1955). On 'exporting' British planning to the world see King (1980). The 1932 British Town and Country Planning Act was influential not only in Australia but throughout the Commonwealth; see Stevens (1955).

15. The international contribution of Australian developments might also be considered. Three starting points are Griffin's Canberra plan, Adelaide and the South Australian parkland town (Williams 1974), and the apparently influential grid of central Melbourne (see Reps 1979:568).
16. Fry (1972:11). Bolton (1981:134) makes a similar point: 'Suburban Australians, like so many latter-day Candides, contented themselves with cultivating their own gardens'. See also: Clarke (1980).
17. See the preface in Sulman (1921:xiii-xiv).
18. Most recently in Bolton (1981:esp. 121-125).
19. For example, Cavanagh (1889), Coane (1891), Keily (1891).
20. But see Walkley (1976). An official history of the profession (Calender 1974 of the Royal Australian Planning Institute:5-9) makes no reference to the Town Planning Institute of Western Australia (1931-1935) although it was Australia's first professional body. Earlier proposals for an institute in New South Wales are reported in The Commonwealth Home (1929), 36, 984:18.
21. C.C. Reade to J.D. Fitzgerald, 13 January 1919, Fitzgerald papers (MS) Inter-State Commission (1917-19).
22. This approach is to some extent anticipated by Spearritt (1972) but more explicitly utilised by Petrow (1978), Burrows (1978) and Barton (1981). A related work on Sir Herbert Gepp is Rayner (1978). Gepp was General Manager of the Electrolytic Zinc Company of Australasia which was responsible for Lutana. A clue to the origin of these theses is Roe (1976). The first general work treating town planning as one of numerous 'progressive' reforms in the early twentieth century appears to be Hyslop (1980). See also Rowse (1978).

REFERENCES

Where references given in the review reoccur in the bibliography a shortened entry has been made followed by a reference to the appropriate section of the bibliography, for example, A2. The full reference will be found in the section indicated.

Alexander, I. (1981) 'Post-War Metropolitan Planning : Goals and Realities', in Troy, P.N., ed., Equity in the City. Sydney, Allen and Unwin.

Artibise, A.F.J. and Stelter, G.A., eds (1979) The Usable Urban Past : Planning and Politics in the Modern Canadian City. Toronto, Macmillan.

Ashworth, W., (1954) The Genesis of Modern British Town Planning.

London, Routledge and Kegan Paul.

Atkinson, A. (1981) 'The Government of Time and Space in 1838.' _The Push fom the Bush : A Bulletin of Social History 1838_, 9:1-19.

Bate, W., (1962) _A History of Brighton_. Melbourne, Melbourne University Press.

Barton, R., (1981) Cadburys .... see A3.

Beed, C.S., (1981) _Melbourne's Development and Planning_. Melbourne, Clewara.

Birrell, J., (1964) Walter Burley Griffin .... see A3.

Bolton, G., (1981) Spoils and Spoilers .... see A2.

Bowman, M., (1976) A History .... see B5.

Broadbent, T.A., (1977) _Planning and Profit in the Urban Economy_. London, Methuen.

Brown, A.J., Sherrard, H.M., with Shaw, J.H., (1969) An Introduction .... see A2.

Brown, A.J. and Sherrard, H.M., (1951) _Town and Country Planning_. Sydney, Angus and Robertson.

Brown, L., (1978) Charles C. Reade .... see A3.

Bryant, P.M., (1967) The Philosophical .... see A3.

Burchett, W., (1979) 'The Planning of East Melbourne' _Victorian Historical Journal_, 50, 2:77-88.

Burrows, A.I., (1978) An Examination .... A3.

Burt, W.O., (1945) 'National, Regional and Local Planning', _Architecture_, 34, 1:151.

Cavanagh, M.F., (1889) 'Street Alignment in Australian Cities', _Building and Engineering Journal of Australia and New Zealand_, 6, 234:155.

Cherry, G.E., (1969) 'The spirit and purpose of town planning - a historical approach'. _Journal of the Town Planning Institute_, 55, 1:12-15.

Cherry, G.E., (1974) _The Evolution of British Town Planning_. Leighton Buzzard, Leonard Hill.

Clark, R.K., (1969a) City Beautiful .... see A3.

Clark, R.K., (1969b) Garden City .... see A3.

Clarke, P., (1980) Queensland Town Planning : some problems and solutions, Unpublished M.U.R.P. Thesis, University of Queensland, Brisbane.

Claval, P., Forester, J. and Goldsmith, W., (1981) _Urban and Regional Planning in an Age of Austerity_. New York, Pergamon.

Coane, J.M., (1891) 'Sanitary subdivision of Land', _Transactions and Proceedings of the Victorian Institute of Surveyors_, 3:85-87. (read April 1889).

Collins, G.R. and Collins, C.C., (1965) _Camillo Sitte and the Birth of city planning_. New York, Randon House.

Colman, J., (1968) Post-war .... see B2.

Connell, R.W. and Irving, T.H., (1980) _Class Structure in Australian History : Documents, narrative and argument_. Melbourne, Longman Cheshire.

Cox, D.N., (1968) A Study .... see B3.

Crow, V., (1978) Haberfield .... see A3.

Davey, G.M., (1969) Town Planning in Ballarat : 1850 - 1865, Unpublished BA Thesis, Department of History, Monash University, Melbourne.
Davison, G., (1979) 'Australian Urban History : A Progress Report' Urban History Yearbook 1979, :100-109. Leicester, Leicester University Press.
Docherty, J.C., (1977) The Second City .... see B2.
Fagence, M., (1979) 'The Political Mileu of Planning - A Case Study : planning, political style and citizen participation in Brisbane 1961 - 1975', Research Paper No 2, Department of Town and Regional Planning, University of Queensland, Brisbane.
Feiss, C., (1943) 'History and the Modern Planner', Journal of the Society of Architectural Historians, 3, 1-2:8.
Fitzgerald, J.D., (1914a) 'Town Planning and City Beautification', Lone Hand, 1, 6:389-392, 448.
Fitzgerald, J.D., (1914b) The Science of Town Planning : its universal application. Sydney, Town Planning Association.
Fitzgerald, J.D., (1918) The Metropolitan Problems of Sydney. Sydney, Government Printer.
Fitzgerald Papers (M.S.) Manuscript, MS.Q.259, Dixson Library, State Library of New South Wales.
Fletcher, P.G., (1977) The Evolution .... see B3.
Freeland, J.M., (1971) The Making of a Profession : a history of the growth and work of the Architectural Institutes in Australia. Sydney, Angus and Robertson.
Freeland, J.M., (1972) People in Cities .... see A2.
Freestone, R., (1981) Australian .... see A3.
Fry, E.C., (1972) 'Growth of an Australian Metropolis', in R.S. Parker, and P.N. Troy, eds, The Politics of Urban Growth. Canberra, Australian National University.
Gawler, J.S., (1963) A Roof .... see B5.
George, K.R., (1971) Henry Ferdinand Halloran .... see A3.
Gibbons, R., (1972) The Development .... see B2.
Gibbons, R., (1976) 'Finance and Planning : the Sydney Water Board and the 1959 Green Belt releases', Australian Journal of Public Administration, 35, 2:147-159.
Gibbons, R., (1980) Improving Sydney .... see B2.
Glasscock, J.T.C., (1967) Colonial Subdivision, Unpublished Dip. U.R.P. Thesis, Queensland Institute of Technology, Brisbane.
Grubb, W.M., (1976a) A History ... see B5.
Grubb, W.M., (1976b) Planning .... see B5.
Hancock, J.L., (1972) 'History and the American Planning Profession : introduction to a new biographical series. Institute of Planners, 38, 5:274-275.
Haratsis, B., (1979) Deja - vu .... see A1.
Harris, C.P. and Dixon, K.E., (1978) Regional Planning in New South Wales and Victoria since 1944 with special reference to the Albury - Wodonga Growth Centre, Canberra, Centre for Research on Federal Financial Relations.
Harrison, P., (1971) 'Planning Sydney :Twenty Five years on', Royal Australian Planning Institute Journal, 9,4:122-129.
Harrison, P., (1973) Walter Burley Griffin .... see A3.

Harrison, P., (1974) 'Planning the metropolitan areas', in I.H. Burnley, ed., Urbanization in Australia : the post war experience, Cambridge, Cambridge University Press.
Harrison, P., (1977a) 'Alfred John Brown : 1893-1976', Royal Australian Planning Institute Journal, 15,2:48-49.
Harrison, P., (1977b) 'Urban Planning' Australian Encyclopedia, 6:184, Grolier Society of Australia.
Harrison, P., (1978) City Planning .... see A2.
Harvey, A., (1964) Walter Burley Griffin .... see A3.
Henry, F.S., (1955) Colonel Light Gardens .... see A3.
Historical Records of Australia, (1914) Series 1, Volume 1:48.
Hugo-Brunt, M., (1975) 'British Colonial Heritage in Planning and Architecture as a background to Australian Settlement', Planner, 15, 4:24-38.
Humphrey, G., (1968) Sir John Sulman .... see A3.
Hyslop, A., (1980) The Social Reform .... see B5.
Inter-State Commission (1917-19) Report No 12, Rents, Commonwealth Parliamentary Paper No 138, 5:449.
Irvine, R.F., (1914) Town Planning : what it means and what it demands, Sydney, Town Planning Association.
Irvine, R.F., (1915) 'National Organisation and National Efficiency', in R.F. Irvine and others, National Efficiency : a series of lectures. Melbourne, Victorian Railways Printing Branch.
Irvine, R.F., (1916) 'Do we want town planning', The Salon, 7, 5:113-118.
Jeans, D.N., (1965) 'Town Planning in New South Wales 1829-1842', Australian Planning Institute Journal, 3, 6:191-196.
Jeans, D.N., (1967) 'Territorial Divisions and the Locations of Towns in New South Wales 1826-1842', Australian Geographer, 10:243-255.
Jeans, D.N., (1981) 'Official Town Founding Procedures in New South Wales 1828-1842', Journal of the Royal Australian Historical Society, 67, 3:227-237.
Johnson, D.L., (1971) Castlecrag .... see A3.
Johnson, D.L., (1973) Walter Burley Griffin .... see A3.
Johnson, D.L., (1977) The Architecture .... see A3.
Johnson, D.L., (1980) Australian Architecture 1901-1951 : Sources of Modernism. Sydney, Sydney University Press.
Johnson, D.L., (1981) Walter Burley Griffin .... see A3.
Johnson, T., (1976) The Development .... see A3.
Jones, D., (1977) Walter Liberty Vernon .... see A3.
Keily, J., (1891) 'Study in Unity of Design in Planning New Towns and New Suburbs', Transactions and Proceedings of the Victorian Institue of Surveyors, 3:88-104 (read September 1889).
King, A.D., (1980) 'Exporting Planning : the colonial and neo-colonial experience', in G.E. Cherry, ed., Shaping an Urban World. London, Mansell.
Larcombe, F.A., (1978) The Advancement .... see B2.
Laverty, J.R., (1954) Development of the Town of Brisbane .... see A1.

Laverty, J.R., (1968) The History .... see B3.
Laverty, J.R., (1971) Town Planning .... see B3.
Ledgar, F., (1976) Planning Theory .... see A3.
Lloyd, C.J., and Troy, P.N., (1978) Federal Intervention .... see A2.
Lloyd, C.J., and Troy, P.N., (1981) Innovation and Reaction : the life and death of the Federal Department of Urban and Regional Development. Sydney, Allen and Unwin.
Logan, T., (1981) Urban .... see B5.
Maiden, H.E., (1960) The History .... see B2.
Matthews, R.S., (1968) Subdivision control .... see B3.
Mayne, A.J.C., (1980a) Disease .... see A1.
McFarlane, B., (1966) Professor Irvine's Economics in Australian Labour History 1913-1933. Canberra, Australian Society for the Study of Labour History.
Migliorino, R., (1976) Development .... see B2.
Murray, R., (1976) Pirouette .... see A3.
Mullins, P., (1979) 'The Struggle against Brisbane's Freeways : 1966-1974', International Journal of Urban and Regional Research, 3,4:542-551.
Oldham, R., and Oldham, J., (1980) George Temple-Poole : Architect of the Golden Years 1885-1892. Nedlands, University of Western Australia Press.
Orchard, L., (1979) Urban Planning in a Market Economy : the example of metropolitan Adelaide, unpublished, Planning Thesis, School of Architecture and Building, South Australian Institute of Technology, Adelaide.
Osborn, F.J., (1952) Review of Town and Country Planning, by Brown and Sherrard, Town and Country Planning, 20, 103:536.
Parliamentary Standing Committee on Public Works, (1917-1918), Inquiry into the proposed scheme for housing commonwealth workmen at Lithgow, Commonwealth Parliamentary Papers, 98:85.
Peek, M.K., (1976) 'Stoneless Bay ....' see A1.
Peisch, M.L., (1964) The Chicago .... see A3.
Petrow, S., (1978) John Daniel Fitzgerald .... see A3.
Planning History Bulletin, (1981) Volume 3, 1:13.
Poole, M., (1979) Town Planning .... see B6.
Powell, A.J., (1969) Planning and Government : a study of the relationship between town planning and government in New South Wales with particular reference to the period bounded by the Local Government (Town and Country Planning) Amendment Act, 1945 and the State Planning Authority Act 1963, Unpublished M.T.C.P. Thesis University of Sydney, Sydney.
Powell, A.J., (1980) Town Planning .... see A2.
Powell, J.M., (1976) Environmental .... see A2.
Price, A.W., (1969) Dacey Garden .... see A3.
Rayner, A.J., (1978) The Ideology of Sir Herbert Gepp : Progressivism and welfare capitalism, unpublished B.A. Thesis, Department of History University of Tasmania.
Report of the Select Committee on the Town Planning and Development Bill (1920) Report, with minutes of proceedings and evidence, South Australian Parliamentary Papers, 69,4.

Reps, J., (1979) Cities of the American West : A History of Frontier Urban Planning. Princeton, Princeton University Press.
Reps, J. (1981) quoted in Planning History Bulletin, 3, 1:13.
Roberts, A.M., (1970) Development ... Annadale .... see A1.
Roberts, A.M., (1978a) City Improvement .... see A1.
Robertson, R.E., (1970) W.E. Bold .... see A3.
Roe, J., (1976) Review of Cities for Sale, Historical Studies, 17, 67:259.
Roe, M., (1976) 'The establishment of the Australian Department of Health : its Background and Significance', Historical Studies, 17, 67:184-185.
Rowse, T., (1978) Australian Liberalism and National Character. Melbourne, Kibble.
Royal Australian Planning Institute Journal (1980) Report on the 1980 Congress, Journal, 18,3.
Royal Commission on the Housing Conditions of the People in the Metropolis, (1917) Victorian Parliamentary Papers, 29.
Rudduck, G., and Grounds, R., (1955) 'The application of the social sciences to town planning in Australia', International Social Science Bulletin, 7, 2:221.
Safran, A., (1977) Legislation 1825-1865 and the Built Form of Melbourne, Research Essay, Dip.T.R.P., University of Melbourne, Melbourne.
Sandercock, L., (1974) Property .... see A2.
Sandercock, L., (1975) Cities .... see A2.
Sandercock, L., (1978) 'A Socialist City in a Capitalist Society : Property Ownership and Urban Reform in Australia', Journal of Australian Political Economy, 3:66-79.
Scott, A.J., (1980) The Urban Land Nexus & the State. London, Pion.
Scott, A.J., & Roweis, S.T., (1977) 'Urban Planning in Theory and Practice', Environment and Planning A, 9, 10:1097-1119.
Scott, M., (1969) American City Planning since 1890. Berkeley, University of California Press.
Skelton, B., Giblin, R., and Pierce, B. (1973) An Industrial .... see A3.
Somers, F.C., (1970) The Life .... see A3.
Spearritt, P., (1972) Consensus .... see B2.
Spearritt, P., (1976) An Urban History .... see B2.
Spearritt, P., (1977) 'John Job Crew Bradfield (1867-1943)' Australian Dictionary of Biography, 7:381-384.
Spearritt, P., (1978) Sydney .... see B2.
Stannage, T., (1977) 'William Ernest Bold (1873-1953)', Australian Dictionary of Biography, 7:335-337.
Stannage, C.T., (1979) The People .... see B6.
Steele, J.G., (1975) Brisbane Town in Convict Days. St Lucia, University of Queensland Press.
Stevens, P.M.H., (1955) 'Planning Legislation in the Colonies', Town and Country Planning, 23, 131:119-123.
Sulman, J., (1890) 'The Laying out of Towns', Report of The Second Meeting for the Advancement of Science, Melbourne, Australasian Association for the Advancement of Science.

Sulman, J., (1921) An Introduction to the study of Town Planning in Australia. Sydney, Government Printer.
Sulman, J., (1927) The Traffic Problems of Sydney. Sydney, John Fairfax.
The Surveyor, (1914), 27, 1:22.
Sutcliffe, A., (1981a) The History of Urban and Regional Planning : an annotated bibliography. London, Mansell.
Sutcliffe, A., (1981b) 'Introduction : the debate on nineteenth century planning' in A. Sutcliffe, ed., The Rise of Modern Town Planning 1800-1914. London, Mansell.
Troy, P.N., (1978) A Fair Price : The Land Commission program 1972-1977. Sydney, Hale and Iremonger.
Van Rooijen, A.V., (1976) George Augustine Taylor .... see A3.
Van Zyl, F.D.W., (1963) 'Adelaide and the "Gridiron' plan in History', Architecture in Australia, 52, 2:98-102.
Walkley, G., (1952) Town and Country .... see B4.
Walkley, G., (1954) 'Pioneer Planning of British Colonies in Malaya and Australia', Proceedings of the Royal Geographical Society of Australasia (South Australia Branch), 55:13-20.
Walkley, G., (1976) The Louis Laybourne Smith .... see A3.
Watts, J., (1976) A History of St Vincent Place, South Melbourne, unpublished Fine Arts Thesis, Melbourne State College, Melbourne.
Webb, M.J., (1968) Planning .... see B6.
Webb, M.J., (1979) Urban .... see B6.
Wheeler, J., (1966) 'Notes on the Moreton Bay Settlement' Australian Planning Institute Journal, 4, 3:76-78.
Wheeler, J., (1967) 'Further notes on the Moreton Bay Settlement', Australian Planning Institute Journal, 5, 3:69-73.
Williams, M., (1966a) 'The Parkland Towns of Australia and New Zealand', Geographical Review, 56,1:67-89.
Williams, M., (1966b) 'Early Town Plans in South Australia', Australian Planning Institute Journal, 4, 2:45-51.
Williams, M., (1974) The Making .... see B4.
Winston, D., (1957) Sydney's .... see B2.
Winston, D., (1976) Nineteenth Century .... see A3.
Yandell, L.V., (1980) The Life .... see A3.

SELECT BIBLIOGRAPHY

There is no comfortable niche for urban planning history in the Bibliography of Urban Studies in Australia (1966 to date) of the Australian Institute of Urban Studies or in any of the three editions of J.H. Shaw's Bibliography of Town and Regional Planning in Australia (1972, 1973/4, 1975). This bibliography is the first attempt to compile a separate guide to the relevant literature. Sutcliffe's The History of Urban and Regional Planning: An annotated bibliography (1981) contains less than 20 Australian references in a total of 1400. This bibliography has nearly 200. It does not purport to be comprehensive but hopefully assembles the most significant contemporary books, monographs, articles and theses covering the period under review. It is not a guide to sources for the study of planning history in

Australia and also excludes historical studies undertaken in Australia but with little or no local content. Titles are organized under a limited number of general headings which are not necessarily mutually exclusive. Two rather subjective criteria for selection of entries were that a work must deal wholly or substantially in part with historical events or precedents to contemporary physical planning in a deliberate and systematic manner, and it must be of the calibre to rate consideration for a hypothetical place in an enlarged edition of Sutcliffe's (1981) bibliography. The references marked with an asterisk are considered to comfortably meet these standards.

A. General and National

A1. Emergence of Urban Planning in Nineteenth Century

Barclay, E. (1971) 'Fevers and Stinks: Some problems of public health in the 1870s and 18802'. Queensland Heritage, 2,4:3-12.

Barrett, B. (1976) The Civic Frontier: the origin of local communities and local government in Victoria. Melbourne, Melbourne University Press.

* Campbell, J. (1970) The Settlement of Melbourne 1851-1893: selected aspects of urban growth. M.A. thesis, University of Melbourne; esp. Ch. 22 ('Problems of the City's Growth in the late Nineteenth Century Environment').

Cannon, M. (1975) Life in the Cities. Melbourne, Nelson.

Davison, G. (1978) The Rise and Fall of Marvellous Melbourne. Melbourne, Melbourne University Press.

Fry, E.C. (1956) The Condition of the Urban Wage Earning Class in Australia in the 1880's. Ph.D. thesis, Australian National University; esp. Part 4 ('Living Conditions').

* Gerathy, G. (1970) The Role of the Sydney Municipal Council in the Development of the Metropolitan Area, 1842-1912. M.A. thesis, University of Sydney.

* Gerathy, G. (1972) 'Sydney Municipality in the 1880s', Journal of the Royal Australian Historical Society, 58, 1:23-34.

* Haratsis, B (1979) 'Deja-vu: town planning in Victoria 1870-1890's', Polis, 6, 2:34-37.

Hunt, S.J. and Bolton, G. (1978) 'Cleansing the Dunghill: water supply and sanitation in Perth 1878-1912', Studies in Western Australian History, 2:1-17.

Kelly, M., ed. (1978) Nineteenth Century Sydney: essays in urban history. Sydney, Sydney University Press.

Larcombe, F.A. (1976) The Stabilization of Local Government in New South Wales 1858-1906. Sydney, Sydney University Press.

Laverty, J.R. (1954) Development of the Town of Brisbane, 1823-1859. B.A. thesis, Department of History, University of Queensland.

Laverty, J.R. (1962) 'The Loan Works Programme of the Brisbane municipal Council 1859-1878', Business Archives and History, 2, 1:55-64.

Lawson, R. (1973) Brisbane in the 1890's: a study of an

Australian urban society. St. Lucia, University of Queensland Press.
McCarty, J. and Schedvin, C.B., eds., (1978) Australian Capital Cities: historical essays. Sydney, Sydeny University Press.
* Mayne, A.J.C. (1980a) Disease, Sanitation and the "Lower Orders": perception and reality in Sydney, 1975-1881. Ph.D. thesis, Australian National University.
Mayne, A.J.C. (1980b) 'City Back Slums in the Land of Promise: some aspects of the 1876 report on overcrowding in Sydney', Labour History, 38:26-39.
Oeser, F. (1966) The Origins of the Melbourne and Metropolitan Board of Works. thesis, Department of Town and Regional Planning, University of Melbourne.
Peek, N.K. (1976) 'Stoneless Bay and Harcourt', Journal of the Canterbury and District Historical Society, Series 2, 8:2-15.
Proudfoot, P. (1975) 'Maritime Influences on the Growth of Sydney', Royal Australian Planning Institute Journal, 13,2:50-57.
Roberts, A.M. (1970) The Development of the Suburb of Annadale, 1876 to 1899. From 'model farm' to 'model township'. B.A. thesis, Department of History, University of Sydney.
*Roberts, A.M. (1978a) City Improvement in Sydney: public policy 188(1900. Ph.D. Thesis, University of Sydney.
Roberts, A.M. (1978b) 'Planning Sydney's Transport 1875-1900' in Kelly, M., ed.,24-36.
Schedvin, C.B. and McCarty, J.W., eds., (1974) Urbanization in Australia: the nineteenth century.Sydney,Sydney University Press.

A2. Developments since 1900
Australian National Commission for UNESCO (1978) Urban Management Processes. Canberra, Australian Government Publishing Service.
Australian Planning Institute (1967) Jubilee Conference. Adelaide: Australian Planning Insitute; papers on the evolution and structure of state planing.
* Bolton, G. (1981) Spoils and Spoilers: Australians make their environment 1788-1980. Sydney, Allen and Unwin; esp. Ch. 11 ('Planners and Improvers').
Bowman, M. (1979) Australian Approaches to Environmental Management: the response of state planning. Environmental Law Reform Group Publication No. 6, Hobart, University of Tasmania.
Brew, K. (1978) The Origins of Planning as Practice. B.T.P. thesis, University of New South Wales.
Brown, A.J. and Sherrard, H.M., with Shaw, J.H. (1969) An Introduction to Town and Country Planning. Sydney, Angus and Robertson, first ed. 1951; esp. Ch. 2 ('Historical Background').
Bunker, R. (1971) Town and Country or City and Region. Melbourne, Melbourne University Press; esp. Ch. 9 (Perth) and Ch. 10 (Sydney).
Fogg, A.S. (1974) Australian Town Planning Law: uniformity and

change. St. Lucia, University of Queensland Press; esp. Ch. 1 ('Town Planning Law: introduction and history').

Fraser, R.D.L. (1972) 'Planning and Government in the Metropolis', Public Administration. 31, 2:123-138.

Freeland, J.M. (1972) 'People in Cities', in Rapoport, A., ed., Australia as Human Setting. Sydney: Angus and Robertson, 99-123.

Harrison, P. (1966) 'City Planning in Australia: what went wrong?' in Wilkes, J., ed., Australian Cities: chaos or planned growth? Australian Institute of Political Science, Proceedings of 32nd Summer School, Sydney, Angus and Robertson, 60-77.

Harrison, P. (1978) 'City Planning' in Scott, P., ed. Australian Cities and Public Policy. Melbourne, Georgian House, 141-173.

Hepburn, J.A. (1960) 'The Impact of Planning on Australian Cities' in Australian Cities in their Regional Environments. Proceedings of the Sixth Australian Planning Congress, Brisbane; reprinted in Town Planning and Local Government Guide (1961), 5,5:194-204.

Kuzelka, R.D. (1969) The Tiger's Back: a report on Australian organizations for metropolitan planning administration. Planning Research Centre, University of Sydney.

* Lloyd, C.J. and Troy, P.N. (1978) 'A history of federal intervention' in Troy, P.N., ed., Federal Power in Australian Cities. Sydney, Hale and Iremonger, 11-37.

Neutze, M. (1978) Australian Urban Policy. Sydney, Allen and Unwin.

Neutze, M. (1981) 'Australian Cities 1938-1988' in Australian 1939-1988: a Bicentennial history bulletin, No. 4, August, 19-26.

*O'Brien, J. (1958) 'Town Planning' in The Australian Encyclopaedia, 8:532-542. Sydney, Angus and Robertson.

'Planning in the Commonwealth: Australia' series (1957) Journal of the Town Planning Institute, 43,9: Canberra (235-238), Victoria (239-242), South Australia (242-245) and Vol. 43, No. 10, November: New South Wales (258-261), Queensland (261-263), Western Australia (263-266).

Powell, A.J. (1966) 'Planning Aims and Political Objectives in Australia', Australian Planning Institute Journal, 4,4:164-169.

Powell, A.J. (1980) 'Town Planning as the art of design and management'. Paper presented to Section 4 (Architecture and Planning), Jubilee ANZAAS Congress.

*Powell, J.M. (1976) Environmental Management in Australia, 1788-1914. Guardians, improvers and profit: an Introductory survey. Melbourne, Oxford University Press; esp. Ch. 12 ('Urban Management and Town Planning in Australia').

*Sandercock, L. (1974) Property, Politics and Power: a history of city planning in Adelaide, Melbourne and Sydney since 1900. Ph.D. thesis, Australian National University.

*Sandercock, L. (1975) Cities for Sale: property, politics and urban planning in Australia. Melbourne, Melbourne University

Press, paperback ed. 1977.
Stretton, H. (1975) Ideas for Australian Cities. Melbourne, Georgian House, first ed. 1970.
Winston, D. (1953) 'Planning in Australia', Report of Proceedings Town and Country Planning Summer School (Bristol), Town Planning Institute, 43-55.

A3. Planners, Planning Ideas and Planned Communities

Barton, R. (1981) Cadburys at Claremont: an Antipodean Bournville? B.A. thesis, Department of History, University of Tasmania.
Bennett, A. (1977) Yallorn: a town built to go. Dip. T.R.P. thesis, University of Melbourne.
Birrell, J. (1964) Walter Burley Griffin. St. Lucia, University of Queensland Press.
Brown, A.G. (1980) 'Landmark Planners on the Queensland Scene or: architects, engineers and surveyors', Planner, 20, 4:59-64.
Brown, L. (1978) Charles C. Reade: a biography. B. Arch. thesis, University of New South Wales.
Bryant, P.M. (1967) The philosophical evolution of Castlecrag and the realization of a concept. B. Arch. thesis, University of New South Wales.
*Burrows, A.I. (1978) An examination of the town planning idea of the garden suburb, and its application to industrial management within the cooperative movement in the early 1920's at the Electrolytic Zinc Company, Risdon, Tasmania. B.A. thesis, Department of History, University of Tasmania.
*Clark, R.K. (1969a) 'The City Beautiful: promise and reality', Architect, 10, 2:40-44.
*Clark, R.K. (1969b) 'The Garden City Movement and Western Australia', Architect, 10, 4:25-32.
Crow, V. (1978) Haberfield: the development of its character. Scrapbook series No. 4, Ashfield and District Historical Society
Freestone, R. (1981) 'Australian responses to the garden city idea'. Paper presented to Section 21 (Geographical Sciences), 52nd ANZAAS Congress, Brisbane.
George, K.R. (1971) Henry Ferdinand Halloran: Town Planner and Proprietor of the Canberra Freehold Estates. B. Arch. thesis, University of Sydney.
*Harrison, P. (1973) Walter Burley Griffin: Landscape Architect. M. Arch. thesis, University of New South Wales.
Harvey, A. (1964) Walter Burley Griffin and Castlecrag. B. Arch. thesis, University of New South Wales.
*Henry, F.S. (1955) Colonel Light Gardens: a study of the garden suburb in Adelaide, South Australia. M.A. thesis, University of Sydney.
Humphrey, G. (1968) Sir John Sulman. Historical research thesis, B. Arch., University of New South Wales.
Johns, G. (1978) Building a Suburb: the Peter Lalor Home Building Cooperative Society. Occasional paper 1, 2, Melbourne State College.
*Johnson, D.L. (1971) 'Castlecrag: a physical and social planning

experiment', Prairie School Review, 8, 3:5-13.
*Johnson, D.L. (1973) 'Walter Burley Griffin: an expatriate planner at Canberra', Journal of the American Institute of Planners, 39, 5:326-336.
Johnson, D.L. (1977) The Architecture of Walter Burley Griffin. Melbourne, Macmillan.
Johnson, D.L. (1981) 'Walter Burley Griffin 1876-1937' in Tanner, H., ed. Architects of Australia. Melbourne, Macmillan, 107-118.
Johnson, T. (1976) The development of Haberfield as a "Garden Suburb": a study of the Green City Movement and its influence on Haberfield, with further discussion of the Federation architecture of the estate. Advanced study report, Department of Architecture, University of Sydney.
Jones, D. (1977) Walter Liberty Vernon (Architect) 1846-1914. B. Arch. thesis, University of New South Wales, 2 vols.
Jones, M.A. (1972) Housing and Poverty in Australia. Melbourne: Melbourne University Press; esp. Ch. 5 ('Slum Clearance') and Ch. 6 ('Area Development').
*Ledgar, F. (1976) 'Planning theory in the twentieth century: a story of successive imports' in Seddon, G. and Davis, M., eds. Man and Landscape in Australia: towards an ecological vision. Australian UNESCO Committee for Man and the Biosphere. Publication No. 2, Canberra, Australian Government Publishing Service, 195-215.
McNamara, T. (1970) Dacey Garden Suburb. Architecture V Research report, Department of Architecture, University of Sydney.
Murray, R. (1976) Pirouette and Pose: a biography of Florence Mary Taylor. Historical research thesis, B. Arch., University of New South Wales, 3 vols.
*Peisch, M.L. (1964) The Chicago School of Architecture: early followers of Sullivan and Wright. New York: Random House; esp. Ch. 7 ('Griffin and the Plan for Canberra').
*Petrow, S. (1978) John Daniel Fitzgerald: the life of an Australian Progressive. B.A. thesis, Department of History, University of Tasmania.
*Price, A.W. (1969) Dacey Garden Suburb. B. Arch. thesis, University of New South Wales.
*Robertson, R.E. (1970) W.E. Bold C.B.E. An analysis of the administrative contribution, community interest and social influence of W.E. Bold, Town Clerk of the City of Perth (Western Australia) 1901-1944. M.A. thesis, University of Western Australia.
Robertson, R.E. (1979) 'W.E. Bold: the town clerk' in Hunt, L., ed., Westralian Portraits. Nedlands: University of Western Australia Press, 175-178.
*Selwood, J. (1979) Residential Development Processes in Perth: two case studies. Geowest 14: Occasional papers of the Department of Geography, University of Western Australia; esp. 30-85 ('City of Perth's Endowment lands').
Shaw, J.H. (1962) 'Dacey Garden Estate'. Technology, 7,3:79-82.
*Shelton, B., Giblin, R. and Pierce, B. (1973) 'An Industrial

Garden City at Claremont, Tasmania', Archetype, 3,4:27-29; reprinted in Forma (1976), 4,1:44-49.
Somers, F.C. (1970) The Life and Works of John Bede Barlow. B. Arch. thesis, University of New South Wales, 5 vols.
Spearritt, P. (1973a) 'The Australian Slum Stigma', Australian and New Zealand Journal of Sociology, 9, 2:41-45.
Taylor, K.J. (1960) Australian Housing Policy and Programmes up to the Second World War. M.A. thesis, University of Sydney.
*Tregenza, J. (1981) 'Charles Reade, 1880-1933: Town Planning Missionary'. Journal of the Historical Society of South Australia, 9:54-79.
Vanags, J. (1979) Walter Burley Griffin and the Implementation of the Canberra Plan. B.A. thesis, Department of History, University of Melbourne.
Van Rooijen, A.V. (1976) George Augustine Taylor, Historical research thesis, B. Arch., University of New South Wales.
*Veale, W.C.D. (1967) 'Charles Compton Reade, First Town Planner - South Australia' in Australian Planning Institute, Jubilee Conference. Adelaide, Australian Planning Institute, 75-79.
Walkley, G. (1976) The Louis Laybourne Smith School of Architecture and Building, South Australian Institute of Technology: A History 1906-1976. Adelaide; esp. 31-37 ('Town Planning').
Wilson, G. (1973) 'Sunshine: a significant development in the Maribyrnong Valley', Archetype, 3, 2:18-19.
*Winston, D. (1976) 'Nineteenth century sources of twentieth century theories: 1800-1939' in Seddon, G. and Davis, M., eds, Man and Landscape in Australia: towards an ecological vision. Australian UNESCO Committee for Man and the Biosphere, Publication No. 2, Canberra, Australian Government Publishing Service, 187-194.
*Yandell, L.V. (1980) The life and works of Dr. J.J.C. Bradfield, Ph.D. thesis, University of New South Wales.

B. State and City

B1. Australian Capital Territory

Atkins, R. (1978) The Government of the Australian Capital Territory. St. Lucia: University of Queensland Press.
Bartholomeusz, C.A. (1969) Canberra, 1910-25: a case study of maladministration in Australia. B.A. thesis, Flinders University of South Australia.
*Benson, A. (1979) 'Canberra design competition of 1911: contender for second place', Canberra Historical Journal, N.S., 3:8-13.
*Birtles, T.G. (1980) An Introduction to the Planning and Development of the Australian Capital Territory. Part 1: Planning for 25,000 people. Canberra, Canberra College of Advanced Education.
Brennan, F. (1971) Canberra in Crisis: a history of land tenure and leasehold administration. Canberra, Dalton.
Brown, A.J. (1952) 'Some Notes on the Plan of Canberra, Federal

Capital of Australia', Town Planning Review, 23, 2:163-165.
Daley, C.S. (1951) 'Canberra and its development', Record of Proceedings: 1951 Federal Congress on Regional and Town Planning. Canberra, 5-11.
Gibson, T.R.S. (1956) 'Canberra's Layout and Development (1911-55)', Journal of the Institution of Engineers Australia, 28, 5:187-198.
Harrison, P. (1964) 'Canberra - the act of faith', Architecture in Australia, 53, 3:76-81.
Harrison, P. (1980) 'Government development in Canberra: ratbag rhetoric, tree planting, ministerial afterthoughts ...', Royal Australian Planning Institute Journal, 18, 3:78-82.
Hodgson, J.D. (1968) Planning the National Capital, 1910-1917. B.A. thesis, Department of History, Australian National University.
*Johnson, D.L. (1980) Canberra and Walter Burley Griffin: a bibliography of 1876 to 1976 and a guide to published sources. Melbourne: Oxford University Press; also published in Vance Bibliographies (1979), Architecture series A-97.
Knight, J. (1981) 'Canberra: product of antagonism', Canberra Historical Journal, N.S., 7:35-40.
Linge, G.J.R. (1975) Canberra: Site and City. Canberra, Australian National University Press.
Pegrum, R.M. (1977) The Politics of Planning: how Australia chose its Federal Capital, M.T.C.P. thesis, University of Sydney.
Priddle, R.A. (1964) 'History and Development of Australia's National Capital'. Journal of the Institution of Engineers Australia, 36, 2:41-46.
*Taylor, G. (1974) The Federal Capital Commission 1925-1930. M.A. thesis, Australian National University.
White, H.L., ed., (1954) Canberra: a nation's capital. Sydney, Angus and Robertson.
Wigmore, L. (1972) The Long View: a history of Canberra, Australia's national capital. Canberra, Dalton, 1st ed. 1963.

B2. New South Wales

Briggs, P. (1978) Public Housing in Sydney: past, present and future. Ph.D. thesis, University of New South Wales.
*Colman, J. (1968) Post-War Planning in the County of Cumberland, New South Wales: a review of local government planning activity in the Sydney metropolitan region since 1945. Planning Research Centre, University of Sydney.
'Cumberland County Council 1945-1964' Retrospect (1964) Australian Planning Institute Journal, 3, 1:3-24.
Docherty, J.C. (1977) The Second City: social and urban change in Newcastle, New South Wales, 1900-c.1929. Ph.D. thesis, Australian National University; esp. Ch. 4 ('Newcastle Unfolding').
*Gibbons, R. (1972) The Development of Sydney: the Improvement of Sydney Commission 1908-09. B.Ec. thesis, Department of Economic History, University of Sydney.
*Gibbons, R. (1978) Transport Administration and Planning in Sydney: co-ordinative influences in a multi-organisational policy

field. M.Ec. thesis, University of Sydney; esp. Part 1 ('The Historical Legacy 1900-1945').
*Gibbons, R. (1980) 'Improving Sydney 1908-1909' in Roe, J., ed., 120-133.
Harrison, P.F. (1972) 'Planning the Metropolis - a case study' in Parker, R.S. and Troy, P.N., eds., The Politics of Urban Growth. Canberra: Australian National University Press, 61-99.
*Larcombe, F.A. (1978) The Advancement of Local Government in New South Wales, 1906 to the Present. Sydney, Sydney University Press; esp. 266-274 ('Town Planning Counties') and 338-343 ('Planning').
Maiden, H.E. (1960) The History of Local Government in New South Wales. Sydney, Angus and Robertson.
Migliorino, R. (1976) Development of Town Planning in New South Wales. B. Arch. thesis, University of Sydney.
Miller, J.D.B. (1954) 'Greater Sydney 1892-1952', Public Administration, N.S., 13, June:110-122 and 13, September: 192-200.
Roe, J., ed. (1980) Twentieth Century Sydney: studies in urban and social history. Sydney, Hale and Iremonger; includes papers on parks, housing as well as transport.
*Spearritt, P. (1972) The Consensus Politics of Physical Planning in Sydney: case studies from the interwar years. B.A. thesis, Department of Government, University of Sydney.
*Spearritt, P. (1973b) Selected Writings of Sydney Planning Advocates, 1900-1947: a preliminary bibliography. Canberra: Metropolitan Research Trust/Sydney: Australian Political Studies Association, Monograph No. 13.
Spearritt, P. (1974) 'Sydney's "Slums": middle class reformers and the Labor response', Labour History, 26:65-81.
Spearritt, P. (1976) An urban history of Sydney 1920-1950. Ph.D. thesis, Australian National University.
Spearritt, P. (1978) Sydney since the Twenties. Sydney, Hale and Iremonger.
Wilcox, M.R. (1968) 'Regulation of Land Use: an historical approach', Valuer, 20, 3:186-188.
*Winston, D. (1957) Sydney's Great Experiment: the progress of the Cumberland County Plan. Sydney: Angus and Robertson.
Winston, D. (1958) 'Background to Town Planning in Australia', Shire and Municipal Record, 51, 8:655-661.

B3. Queensland

Cass, T.R. (1979) Regional Planning in Australia: retrospect and prospect, M.U.R.P. thesis, University of Queensland.
*Cox, D.N. (1968) A Study of the Formation of the Brisbane Town Plan, M.A. thesis, University of Queensland.
*Fletcher, P. G. (1977) The Evolution of Brisbane's Town Plan 1926-1976, Dip. U.R.P. thesis, Queensland Institute of Technology.
Greenwood, G. and Laverty, J. (1959) Brisbane 1859-1959: a history of local government. Brisbane, Brisbane City Council.
Laverty, J.R. (1968) The History of Municipal Government in Brisbane 1859-1925: a study of the development of metropolitan

government in a context of urban expansion, Ph.D. thesis, University of Queensland.

*Laverty, J.R. (1971) 'Town Planning in Brisbane 1842-1925', *Royal Australian Planning Institute Journal*, 9, 1:19-26.

Laverty, J.R. (1972) 'Greater Brisbane: a response to problems of metropolitan government', *Australian Journal of Politics and History*, 18, 1:34-51.

Laverty, J.R. (1978) 'Greater Brisbane in Retrospect' in *Brisbane Retrospect Seminar: eight aspects of Brisbane history*. Brisbane, Library Board of Queensland, 24-45.

Matthews, R.S. (1968) Subdivision Control in Brisbane 1859-1965, Dip. T.R.P. thesis, University of Queensland.

Toms, K.N. (1974) *Urban Government, Politics and Planning: a study of the administration of the Brisbane Town Plan*. Brisbane, Queensland Regional Group of the Royal Institute of Public Administration.

Wheeler, J. (1968) 'Fifty Years of Planning', *Planner*. 8, 5:16-23.

B4. South Australia

*Bennett, C.J.C. (1976) The Role of Local Government in the Administration and Development of Town Planning in South Australia, M.T. P. thesis, University of Adelaide.

Cain, F.M. (1964) The State and Housing in South Australia 1920 to 1930, B.A. thesis, University of Adelaide.

Stretton, H. (1978) 'The South Australian Tradition of Radical Land Administration', *Australian Surveyor*, 29, 2:89-92.

*Tan, Z.H. (1978) Planning and Politics: a case study of Adelaide (1938-1968). M.U.R.P. thesis, University of Adelaide.

*Walkley, G. (1952) 'Town and Country Planning in South Australia', *Proceedings of the Royal Geographical Society of Australasia (S.A. Branch)*, 53:79-92.

Williams, M. (1974) *The Making of the South Australian Landscape: a study in the historical geography of Australia*. London, Academic Press; esp. Ch. 9 ('The Making of Adelaide').

B5. Victoria

*Bowman, M. (1976) 'A History of Town Planning in Melbourne', *Polis*, 3, 2:1-4.

Gawler, J.S. (1963) *A Roof Over My Head*. Melbourne, Lothian.

*Grubb, W.M. (1976a) A History of Town Planning in Victoria 1910-1944. M.T.R.P. thesis, University of Melbourne.

*Grubb, W.M. (1976b) 'Planning in Victoria 1910-1943', *Polis*, 3, 2:5-9.

*Hyslop, A. (1980) The Social Reform Movement in Melbourne: 1890 to 1914, Ph.D. thesis, Latrobe University; esp. Ch. 11 ('Environmental and Health Measures for Town and Country').

Ledgar, F.W. (1960) 'Town Planning in Victoria with particular reference to the aim and scope of the Melbourne Metropolitan Planning Scheme', *Melbourne University Law Review*, 2, May: 283-302.

Logan, T. (1981) *Urban and Regional Planning in Victoria*. Melbourne, Shillington House.

Paterson, J., Yencken, D. and Gunn, G. (1976) A Mansion or No House: a report for UDIA on consequences of planning standards and their impact on land and housing. Melbourne: Urban Development Insitute of Australia; esp. Ch. 2 ('How Standards Legislation was Introduced').

Russell, E.W. (1972) The Slum Abolition Movement in Victoria 1933-37. Melbourne, Hornet Publications.

Turner, I.A.H. (1967) 'The Growth of Melbourne: an historical account' in Troy, P.N., ed. Urban Redevelopment in Australia. Canberra: Urban Research Unit, Australian National University, 19-48; reprinted in McCarty, J. and Schedvin, C.B., eds., (1978).

Vogt, D. (1975) A History of Conflict: Melbourne town planning in the 1920's, B.A. thesis, Department of History, University of Melbourne.

Wilson, G. and Sands, P. (1981) Building a City: 100 years of Melbourne architecture. Melbourne, Oxford University Press; esp. Ch. 8 ('Melbourne as a Town Plan').

B6. Western Australia

Boas, H. (1956) 'The Evolution of Town Planning in Western Australia'. Paper presented to Fourth Australian Planning Congress, Perth.

Carr, D. (1979) 'Metropolitan Design' in Gentilli, J., ed., Western Landscapes. Nedlands, University of Western Australia Press, 383-399.

Georgiou, J. (1979) 'The Metropolitan Region' in Pitt Morison, M. and White, J., eds. Western Towns and Buildings. Nedlands, University of Western Australia Press, 247-265.

Hilford, C.H. (1972) Forrest Place, Perth: its origins, present function and future role, M. Arch. thesis, University of Western Australia.

Logan, L.A. (1967) Town Planning: Fifty Years. Information Booklet No. 1, Western Australian Town Planning Department.

*McLeod, P.B. and Ellis, J.R. (1979) 'Planning for the Perth Region: a brief history and commentary', Economic Activity, 22, 3:12-18.

*Poole, M. (1979) 'Town Planning' in Pitt Morison, M. and White, J., eds., Western Towns and Buildings. Nedlands, University of Western Australia Press, 220-232.

Stannage, C.T. (1979) The People of Perth: a social history of Western Australia's capital city. Perth: Perth City Council; esp. Ch. 4 ('Forests in the City - Perth 1884 to 1918').

*Webb, M.J. (1968) 'Planning and Development in Metropolitan Perth to 1953' in The City and Region of Perth: a case study of the capital city of W.A. Papers and proceedings of the Tenth Congress of the Australian Planning Institute, Perth.

*Webb, M.J. (1979) 'Urban Expansion, Town Improvement and the Beginnings of Town Planning in Metropolitan Perth' in Gentilli, J., ed., Western Landscapes. Nedlands, University of Western Australia Press, 359-382.

Policy review
Urban affairs in South Australia 1981

Ray Bunker

General conditions in South Australia during 1981 could have been better. In March and September the national unemployment rate of those seeking full-time work was 5.8%, as estimated by the Australian Bureau of Statistics. In both these months, South Australia's rate was the highest of all states and had deteriorated from 7.7% to 8.5% in the six-month period. Similarly, population growth continued slowly. Departures from the state exceeded arrivals from mid 1978 to the last available figures showing conditions at the end of 1980.

Economic development

Unlike most of the other states, the growth in exploitation of minerals and sources of energy was not strongly represented in South Australia in 1981. The open-cut coal mine at Leigh Creek is being enlarged and expanded to serve a new power station being built at Port Augusta. The existing town covers part of the coal needed and a new town at Leigh Creek South has been built to which workers and their families began to transfer late in the year. Approval was given for processing and shipping facilities to be built at Stony Point, near Whyalla to process liquids from the Cooper Basin oil and gas fields. The approval included the necessary pipeline from the Basin which was also capable of taking extra ethane, if needed, to provide feedstock for a possible petrochemical plant - a revival of the Redcliffs proposal of the 1970s. Another natural resource - larger and longer-term - is the Roxby Downs deposit. This continues a mixture of minerals, one of the most important of which is uranium. There is considerable political controvesy about the venture. Some technical considerations and issues about the provision of infrastructure also remain to be resolved.

In manufacturing industry, no substantial changes occurred. There are signs that the whitegoods industry in South Australia will reconstruct itself effectively to meet the eventual lowering of tariffs proposed by the Industries Assistance Commission. Another significant South Australian industry - car-making - waited somewhat nervously to see what the IAC proposed for it and what the Federal Government would do. There were demonstrations by car workers.

The construction industry remained in a depressed state. Early 1981 showed the number of dwellings under construction at the lowest level for many years : they had fallen from 4,778 in mid 1978 to 3,060 in March 1981.

Public Finance

Commonwealth payments to South Australia contained in the Federal budget of August showed a decline in real money terms for 1981/82 compared with the previous year. Significantly, while general and recurrent purpose funding increased, general purpose capital funding was held at the same money level and payments for specific capital spending purposes actually declined even in these terms. This situation was not helped in the state budget presented a month later.

While the budget proposals for 1981/82 forecast a small deficit of $3M, this comprised a $4M loss in the operation of current activities, compensated by a transfer of moneys provided for capital spending purposes. These gloomy trends in public finance were not helped by the release in June of the Report on Tax Sharing Entitlements 1981 by the Commonwealth Grants Commission. The commission's calculations indicated that the personal income tax sharing entitlement of South Australia should be reduced from the present differential of 153% to 132% per capita payment compared with the Victoria base of 100%.

Urban Development

In 1980 the State Government reached an agreement with the Federal Government to terminate the financial arrangements governing the Monarto growth centre. In November 1981 the sale commenced of some of the properties acquired to build the new city years ago. Some of the wooded parts of the site are to be used as a park. In the same month the Premier announced he had also reached agreement with Canberra to reconstruct and repay the State's debt on the Land Commission. The sum to be repaid was half that owing in the middle of the year (including capitalised interest) and repayments were to be spread over a period. Earlier in the year the Government had passed legislation to reduce the powers and role of the Land Commission. In the Urban Land Trust Act the renamed Commission is to have a largely land-banking role, is not to acquire further land without the approval of the Minister, and is to have no land developmental functions apart from subdividing its broadacre holdings for other parties to process further.

Fringe development of metropolitan Adelaide declined with the drop in new dwelling construction. The government was persuaded of the merits of urban consolidation. The activities of the Housing Trust reflected this orientation, but some policies were clearly contrary to compact development. The Valuation of Land (Amendment) Act shocked followers of Henry George. It introduced the concept of 'notional' value. Property used as a principal place of residence or for commercial farming purposes was to be valued at existing and not at potential use where the possibility of change to a higher and/or better use existed. This could be of benefit to rural producers on the fringe of Adelaide. But it did nothing to encourage the infill of vacant areas for market gardening areas within the urban fabric.

Housing

The average rent for a house in Adelaide rose from $54 per week in May 1980 to $65 a year later while vacancy rates fell to very low levels. The difficult circumstances attending shelter were particularly apparent in the activities of the Housing Trust. Applications for rental accommodation rose to a record level of 11,501 in 1980/81. Despite a record number of allocations, the waiting list was over 20,000 in the middle of the year. Nearly half the stock of dwellings for rent (43,652) was occupied by households in receipt of rent reductions.

While Commonwealth funds for public housing fell dramatically, considerable inventiveness was shown in turning to other sources of funds. The State Government increased its funding and substantial sums are being raised by promissory notes, and loans from the State Government Insurance Commission and Superannuation Fund. Other initiatives are being taken in mounting joint ventures with private builders (the "design and construct" programme) private landlords (lease and subsidisation of rents of tenants), local government, housing cooperatives and service organisations. The result of all this is that over $100M will be available for the Trust's capital works programme in 1981/82 and the construction of 1,700 dwellings commenced - 500 more than the previous year's total.

Planning

Significant events here were the amalgamation of the former Departments for the Environment and of Urban and Regional Affairs into a new Department of Environment and Planning. This brought together a number of related functions. This organisational change was reflected in new legislation. A bill designed "to streamline and simplify the way in which planning and development is controlled in South Australia, to make the planning process more flexible and to reflect the wishes of the community", was introduced in June and was allowed to lie on the House of Assembly table while consultation with interested parties took place. A revised bill introduced in November seemed certain to pass through Parliament before Christmas. The bill provides some of the processes and mechanics for better planning, and it is hoped that the substance and style of operation will also improve. A dual occupancy policy was strongly recommended to local councils by the State Planning Authority. Most of them were non-commital or opposed to the policies as presented.

Transport

Four significant events took place here in 1981. Research conducted by or for the Department of Transport went far to establish the costs of providing various kinds of public transport services and some moves were taken to reflect this in pricing. Second, the government decided to drop long-established road-widening plans for many major inner metropolitan arterial roads. Some 60% of the roads previously affected were withdrawn by this action. This decision frees fronting properties from restrictions on sale, purchase and development. In similar vein some land acquired for transportation corridors is to be released and sold. Third, a revised busway plan was announced for the north east corridor linking the city to Tea Tree Gully. The plan uses the German O-Bahn guided busway system, running for most of its route on a bus-only road up the valley of the River Torrens. Finally, Adelaide's airport is to be extended to take limited international flights - the only mainland airport without this capability.

Water supply

South Australian concern for the quality and amount of its water supply achieved national prominence in various ways in 1981. The River Murray Commission is to be given authority to control the quality as well as the quantity of River Murray water. The State Government submitted a programme of works to combat salinity to the Federal Government as a national bicentennial project. And the Engineering and Water Supply Department is considering what pricing policies are appropriate for the supply of water and disposal of sewerage.

Review articles

Australia's cities: in or out of print

Peter Spearritt

> The light was stronger now. You could see the red roofs and the spires of churches and the grey skyline of the city. You could see the dome of the Zoo on the opposite hillcrest, and the great, ghostly arc of the bridge...You were one of the red roofs, and all about you, on this shore and on the opposite shore, from Balgowlah to Parramatta, were your neighbours, the other red roofs...

Thus did Eleanor Dark introduce Sydney in her novel Waterway, first published in 1938. Like many Australian novelists she chose one of Australia's growing cities as the setting for her plot. This interest in urban Australia was not reflected in the universities where architecture students learnt about London and Venice while students of English literature studied just that. Ironically, it took an Englishman, Asa Briggs, to point out that our cities were fit subjects for scholarly scrutiny (1961, 1963). But when he addressed an audience at the University of Sydney on 'Historians and the Study of Cities' in 1960 he was not particularly well received in a university where Australian history courses stopped with the First Fleet. Briggs pointed out to his audience that

> the cities of Australia are at the present time either in the process of drastic transformation or on the eve of it. Consequently, cities are in the news again as they have not been since the great Victorian boom of the 1880s (Briggs, 1960:5).

Briggs was right. Australian cities were in the news in 1960. Real estate agents, speculative builders, home buyers and owner-builders jostled with each other in many an outer suburb, while insurance companies competed to plan the highest downtown office blocks. Judging from the lack of writing about these topics, most of our social scientists at the time were more interested in getting a foothold on the home ownership ladder than they were in analysing it.

The handful of scholarly volumes produced in the 1960s showed promise of a growing awareness of urban Australia, but their print runs were so small and their circulation so restricted that their impact was severely limited (for example, Wilkes, 1966; Troy, 1967). The standard social science text-book fare for that decade were edited volumes like Australian Civilization (Coleman, 1962) where 'cities' did not even rate a chapter.

In 1970 Hugh Stretton's Ideas for Australian Cities burst upon this sleepy world. Stretton claimed on the back cover that 'six commercial publishers' rejected the book which he self-published under the alluring imprint 'An Orphan Book'. In fact, as he later admitted, some would have published it, but not in paperback and not for two dollars, his two main conditions. At a modest $1.95 this very readable book became an overnight bestseller and Stretton had no difficulty in finding a commercial publisher to take over the demand for reprints (Stretton, 1970 and 1975).

Stretton's book, the work of an historian who had turned his attention to contemporary urban life, was historically informed and politically sensitive, though it relied heavily on the South Australian Housing Trust and the National Capital Development Commission as models for future government intervention in urban planning (cf. Jones, 1979). But Adelaide and Canberra are not Sydney or Melbourne and the book's prescriptions for state intervention in those cities were optimistic about the level of egalitarian rationality to be found in political parties and the state bureaucracies.

Despite the success of Stretton's Ideas commercial publishers were even slower than academics to appreciate the importance of the field. The next three books of note, Parker and Troy., The Politics of Urban Growth (1972), Jones' Housing and Poverty (1972) and Sandercock's Cities for Sale (1975) were all published by university presses and inadequately distributed as a result. All had relatively modest sales, with only Sandercock going to a second printing (Sandercock, 1975 and 1977). Like Stretton she assumed that town planners could hold the key to urban reform in Australia. That may have appeared to be a possibility in the boom of the early seventies, but in the ensuing recession it became clear that economic forces would have much more influence on urban form and content (from housing types to job creation and location) than 'planners' ever would. Ironically, the Department of Urban and Regional Development recognised this in its attempt to construct an urban and regional budget in opposition to Treasury's conventional model. One of the reasons for the Department of Urban and Regional Development's demise at the hands of the incoming Liberal government was that the traditional holders of economic expertise - the Treasury - did not wish to have their central bureaucratic power questioned (cf. Lloyd and Troy, 1981:260).

All the aforementioned volumes had, as their central concern, current urban policy and its future direction, though both Jones and Sandercock also studied historical antecedents. Simultaneously with this growing interest in urban policy was a growing interest in urban history. Much of the inspiration for this came in the first instance from Britain, as can be seen in the special issue of the Australian Economic History Review 'Australian Capital Cities in the Nineteenth Century', where the familiar British concerns with service provision and housing tenure predominated. Another British academic, Sean Glynn, produced his Urbanisation in Australian History 1788-1900 in the same year. Most subsequent urban history continued to be preoccupied with the latter half of the 19th century, as in Lawson's 1973 book on Brisbane in the 1890s, Kelly's 1978 edited volume Nineteenth Century Sydney and in Davison's 1978 volume on Melbourne in the 1880s.

Melbourne's urban historians are uncomfortable with suburbanisation since the second world war because it robs them of one of their central themes - the evolution of community. Thus Bate's Brighton (1962) and Lack's Footscray (1980) peter out in the interwar years when job dispersal, increasing car usage and the entry of more married women into the paid workforce cut across more tight-knit suburban patterns. Notions of community have been more successfully employed in studies of country towns and isolated urban settlements (see Bowman, 1981).

Much of the debate in urban history now centres around whether researchers should concentrate on individual and group experience or urban form and structure. In their respective urban histories, Stannage (1979, on Perth) concentrated on the former, Spearritt (1978, on Sydney) on the latter while Davison (1978) oscillated between the two approaches (see Davison, 1979; McCarty, 1980; Stannage, 1982). Bate's Lucky City : The First Generation at Ballarat 1851-1900 (1978) manages to marry both approaches, but whether this can be successfully done with large metropolitan populations as distinct from much smaller urban settlements remains to be seen.

Print and Prosper?

By 1978 two commercial publishers had decided that the example of the presses might be worth following. This gave a vital boost to research and publication in Australian urban studies. That year saw George Allen and Unwin produce its first book under an Australian imprint, Neutze's Urban Development in Australia. This book drew heavily on previous work published by the Australian National University's Urban Research Unit and distributed on a very restricted basis through its own monograph series. In the same year a new Sydney publisher, Hale and Iremonger, produced four volumes in urban studies including Roddewig's Green Bans and Troy's A Fair Price. Both A Fair Price and Federal Power in Australian Cities (Troy, 1978b) reflected on the experience of federal labor governments efforts to initiate urban reform, including a more equitable distribution of housing, land, jobs and services. Lloyd and Troy's Innovation and Reaction, reviewed in this yearbook, takes up these themes, while A Just Society (Troy, 1981a) and Equity in the City (Troy, 1981b) examine the philosophical underpinnings and empirical reality of the distribution of wealth and resources in Australia. The latter two volumes are tentative forays rather than final statements in a developing debate.

Very few books about urban Australia have lived up to the early promise of Stretton's Ideas. Our urban geographers are just as ahistorical and apolitical as they have always been: Burnley's The Australian Urban System (1980) and Logan et.al. Urbanization : The Australian Experience (1981) are remarkable for their lack of use of the now plentiful work by urban historians, work which should have been used to test and illuminate many of their observations. Historians may have prevented the wider dissemination of their work by concentrating on illustrative detail to the exclusion of a necessary analysis of form and structure, though most of them at least read outside their own discipline, which is more than can be said for many geographers and economists. As political scientists have to date offered little to urban studies, the apolitical nature of much writing in the field can perhaps be excused. Australian social scientists are more likely to draw inspiration about how to approach urban politics from the work of Castells, Dunleavy and Saunders (discussed in Chris Paris' review article) than they are from home grown work, although a number of volumes are now being published (for example, Parkin, 1982).

In the 1950s architects and planners made some notable contributions to our understanding of metropolitan development but they have had little to say of late. Robin Boyd's Australia's Home : Its Origins, Builders and Occupiers, first published in 1952, remains the best work on suburban housing, despite thirty years of assiduous research by architects, planners, economists and political economists. A major book drawing together this extensive research - presently hidden away in government reports, state government and council archives and long-forgotten articles - is yet to be written.

Sociologists have attempted to synthesise empirical and theoretical work in Australian urban studies (for example, Kilmartin and Thorns, 1978) without conspicuous success, in part a reflection of the disparate nature of work in that discipline. Connell and Irving's Class Structure in Australian History has produced more hypotheses for Australian urban studies than any other historical or sociological work, but few of them - such as the assertion that 'In the new outer suburbs, working-class families found more physical freedom, but a new economic dependence' (1980:299) have yet been tested.

The best general book on urban policy remains Neutze's Australian Urban Policy (1978), a companion volume to his Urban Development in Australia (1977, revised edition 1982). Taken together these two volumes canvass - from the point of view of an urban economist - the key areas of transport, housing, employment and the role of government in urban development. They have been attacked for being apolitical (for example, Parkin, 1979) and atheoretical, though just what theory means in urban studies is a matter of continuous debate and disagreement. To some it means model testing or neat taxonomies, to others a priori explanations of all aspects of social political and economic life.

In particular aspects of urban studies we are now relatively well-informed. We know a great deal about the distribution of urban populations and dwellings in Australia, but as Nigel Thrift points out in his review of the various atlases published in the last few years this takes only part of the way in understanding urban social processes. We still know remarkably little about how urban economies work and about their internal and external linkages. We know much more about the location of unemployment and the distribution of various types of employment than we do about how and why jobs appear and disappear. We know more about the journey to work (for example, Manning, 1978) than about any other intra-urban trips and we know an increasing amount about the inner city (for example, Kendig, 1978) and outer suburbia (for example, Bryson and Thompson, 1972) but remarkably little about the suburbs in between.

The role of the state in urban Australia has now been a key theme in a number of works, though most of these have not developed or addressed themselves to specific theories of state intervention. Thus we know much more about federal government urban policies and the role of state and local governments (for example, the books by Neutze, Troy and Power et.al.) than we do about the workings of the real estate and property development market. An Australian equivalent of Massey and Catalano's Capital and Land (1978) is sorely needed. Daly's Sydney Boom, Sydney Bust (1982) is a useful beginning, but it is more concerned with developer than consumer finance, and in a city where seventy per cent of all dwellings are owned or being purchased, consumer fiance is a vital ingredient in the urban economy.

Feminists have only recently begun to write about Australia's cities, but their work - as in the case of Allport's article in this issue - has major new insights to offer. The 1976 census income statistics are a rich and largely untapped source for understanding the key factors that make up income distribution - gender, age and occupation - and the implications of that distribution for life in the urban areas (a preliminary attempt is made in Poulsen and Spearritt, 1981).

For budding intellects or jaded scholars seeking new directions, there are a number of exciting possibilities in writing about Australia's cities. Good books are yet to be written on the development, nature and extent of property rights in Australia and of the relations between landlords and tenants. We don't even know who our landlords are. We lack an adequate account of the dynamics of our metropolitan economies and their place in our state, national and international economies, though Stilwell's Economic Crisis : Cities and Regions (1980) is certainly a move in the right direction. Urban histories of all our capital cities in the twentieth century, except Sydney, remain to be written, along with the histories of many of our major urban areas. Perhaps the coming bicentennary will encourage more local councils and state governments to throw open their recent archives to researchers.

Urban studies is obviously here to stay and while not particularly robust, has a growing Australian literature of considerable substance. The opportunities are many and the publishers are now willing, so if you have something worth saying there is a fair chance you can say it in print.

REFERENCES

Bate, W. (1962) A History of Brighton. Melbourne, Melbourne University Press.

Bate, W. (1978) Lucky City : The First Generation at Ballarat 1851-1900. Melbourne, Melbourne University Press.

Boyd, R. (1952) Australia's Home : Its Origins, Builders and Occupiers. Melbourne, Melbourne University Press (reprinted Penguin 1968).

Briggs, A. (1960) Historians and the Study of the City, G.J. Cohen Memorial Lecture, University of Sydney.

Briggs, A. (1961) 'The Sociology of Australian Cities', Outlook August:10-11.

Briggs, A. (1963) Victorian Cities. London, Odhams.

Bowman, M. ed. (1981) Beyond the City : case studies in community structure and development. Melbourne, Longman Cheshire.

Bryson, L. and Thompson, F. (1972) An Australian Newtown. Melbourne, Penguin.

Burnley, I. (1980) The Australian Urban System. Melbourne, Longman Cheshire.

Coleman, P. ed. (1962) Australian Civilization. Melbourne, Cheshire.

Connell, R.W. and Irving, T. (1980) Class Structure in Australian History. Melbourne, Longman Cheshire.

Davison, G. (1978) The Rise and Fall of Marvellous Melbourne. Melbourne, Melbourne University Press.

Davison, G. (1979) 'Australian Urban History : A Progress Report' in Urban History Yearbook:100-109, Leicester, Leicester University Press.

Daly, M.T. (1982) Sydney Boom Sydney Bust : The City and its Property Market 1850-1981. Sydney, Allen and Unwin.

Glynn, S. (1970) Urbanisation in Australian History 1788-1900. Melbourne, Nelson.

Jones, M. (1972) Housing and Poverty in Australia. Melbourne, Melbourne University Press.

Jones, M. (1979) 'Australian Urban Policy', Politics, 14, 2:295-303.

Kelly, M. ed. (1978) Nineteenth Century Sydney. Sydney, Sydney University Press.

Kendig, H. (1978) New Life for Old Suburbs. Sydney, Allen and Unwin.

Kilmartin, L. and Thorns, D. (1978) Cities Unlimited : The Sociology of Urban Development in Australia and New Zealand. Sydney, Allen and Unwin.

Lack, J. (1980) 'Residence, Workplace, Community : Local history in metropolitan Melbourne', Historical Studies. 19, 74:16-40.

Lawson, R. (1973) Brisbane in the 1890s. Brisbane, University of

Queensland Press.
Lloyd, C. and Troy, P.N. (1981) Innovation and Reaction : The Life and Death of the Federal Departement of Urban and Regional Development. Sydney, Allen and Unwin.
Logan, M.I., Whitelaw, J.S. and McKay, J. (1981) Urbanization: The Australian Experience. Melbourne, Shillington House.
McCarty, J.W. (1980) 'Melbourne, Ballarat, Sydney, Perth : the new city histories', Historical Studies. 19, 74:1-15.
Manning, I. (1978) The Journey to Work. Sydney, Allen and Unwin.
Massey, D. and Catalano, A. (1978) Capital and Land : Landownership by Capital in Great Britain. London, Arnold.
Neutze, M. (1977) Urban Development in Australia. Sydney, Allen and Unwin (revised edition 1982).
Neutze, M. (1978) Australian Urban Policy. Sydney, Allen and Unwin.
Parker, R.S. and Troy, P.N. eds., (1972) The Policy of Urban Growth. Canberra, Australian National University Press.
Parkin, A. (1979) 'Cities without politics', Politics. 14, 2:291-294.
Parkin, A. (1982) Governing the Cities. Melbourne, Macmillan.
Power, J., Wettenhall, R. and Halligan, J. eds., (1981) Local Government Systems of Australia. Canberra, Australian Publishing Service.
Poulsen, M. and Spearritt, P. (1981) Sydney : A Social and Political Atlas. Sydney, Allen and Unwin.
Roe, J. ed. (1980) Twentieth Century Sydney. Sydney, Hale and Iremonger.
Roddewig, R.J. (1978) Green Bans. Sydney, Hale and Iremonger.
Sandercock, L. (1975, 1977) Cities for Sale : Property, Politics and Urban Planning in Australia. Melbourne, Melbourne University Press (paperback edition 1977).
Spearritt, P. (1978) Sydney Since the Twenties. Sydney, Hale and Iremonger.
Stannage, T. (1979) The People of Perth. Perth, Carroll.
Stannage, T. (1982) 'Australian Urban History' in Osborne, G. and Mandle, W.F. eds., New History : Studying Australia Today. Sydney, Allen and Unwin.
Stilwell, F. (1980) Economic Crisis : Cities and Regions. Sydney, Pergamon.
Stretton, H. (1970) Ideas for Australian Cities. Adelaide, Orphan Book (reprinted for Georgian House).
Stretton, H. (1975) Ideas for Australian Cities. 2nd ed., Adelaide, Georgian House.
Troy, P.N. ed. (1967) Urban Redevelopment in Australia. Canberra, Urban Research Unit, Australian National University.
Troy, P.N. ed. (1978a) Federal Power in Australia's Cities. Sydney, Hale and Iremonger.
Troy, P.N. ed. (1978b) A Fair Price : The Land Commission Program 1972-1977. Sydney, Hale and Iremonger.
Troy, P.N. ed. (1981a) A Just Society?. Sydney, Allen and Unwin.
Troy, P.N. ed. (1981b) Equity in the City. Sydney, Allen and Unwin.
Wilkes, J. ed. (1966) Australia's Cities : Chaos or Planned Growth?. Sydney, Angus and Robertson.

Urban politics: where to now?

Chris Paris

Patrick Dunleavy, Urban Political Analysis : The Politics of Collective Consumption, Macmillan, London, 1980, pp. 196, rrp. £4.95.
Peter Saunders, Urban Politics : A Sociological Interpretation, Penguin Books, Harmondsworth, 1980, pp. 383, rrp. $9.95 paper.
John Power, Roger Wettenhall and John Halligan, eds., Local Government Systems in Australia, Advisory Council for Inter-Governmental Relations Information Paper No 7, Australian Government Publishing Service, Canberra, 1981, pp. 830, rrp. $25.60 paper.

Australian social scientists traditionally have not displayed much enthusiasm for the study of urban politics. During the 1970s, however, it was apparent that the development of and changes in Australian cities were political issues. From the Department of Urban and Regional Development to the sacking of Melbourne City Council, from green bans to anti-freeway movements, or from inner city tenants' protests to the problem of the mortgage battler - 'urban' issues have periodically waxed and waned on the political agenda.

In virtually all other advanced capitalist societies, too, intellectuals and politicians identified growing 'urban crises'. With the growth of the so-called new urban sociology, particularly through the writing of Manuel Castells, a case was made strongly for the growing significance of urban politics. The concluding chapter to City, Class and Power (Castells, 1978) started with the claim that our understanding of what is meant by the 'urban' in advanced capitalist societies has been transformed, both in theory and practice. Moreover:

> ...urban politics are at the centre of the urbanisation process, ...they play an increasingly important part in the whole of the political process as such (Castells, 1978:174).

Castells, however, used the term 'urban politics' in a specific and controversial way. Indeed, a survey of his many published works shows that diverse and at times contradictory meanings were given both explicitly and implicitly to the term. His suggestion that urban politics are universally of increasingly significance is also highly questionable, particularly from an Australia perspective. What is specifically 'urban' about urban politics, how significant are the issues which Castells raises, and how useful are his ideas in Australia? The first two questions are raised by two British books which critically reflect on Castells' work and go on to suggest alternative definitions of urban politics (Dunleavy, 1980; Saunders, 1979). In addition, we can go some way towards evaluating the relevance to Australia of overseas theoretical developments in the light of a major recent study of local government in Australia (Power, Wettenhall and Halligan, 1981).

DUNLEAVY AND SAUNDERS ON 'URBAN POLITICS'

Saunders' first book (1979) followed an earlier Ph.D. on community power based on Croydon, an outer London borough. It comprises an empirical study reported in part 2 which was completed before most of the theoretical work in part 1 was undertaken. In this review I am focussing almost entirely on the latter. Saunders, in his theoretical discussions, was particularly interested in the way that local decision-making affects the distribution of costs and benefits of public policies and the role of bureaucratic 'gatekeepers'. This followed his assumption that political power derives from elected position or bureaucratic office. He adopted an explicitly weberian approach using a 'housing class' analysis which maintained that ownership of domestic property has led to the creation of a 'middle' property class. This 'new middle class' (Carchedi, 1975) was seen to pose a problem for marxist theory:

> ...either home ownership must in some ways be included in an analysis of class position, or it must be recognised that political struggles involve groups which cannot be analysed in class terms (Saunders, 1979:100).

Hence, Saunders (1979:100) argued, following Hirst (1979), that there was a '<u>necessary</u> non-correspondence between economic class categories and political action' (loc cit). Saunders found Castells' (1977a, 1977b) approach to 'collective consumption' and 'urban social movements' unconvincing, particularly rejecting the notion that 'a common political situation in the urban system' can overcome economic class divisions. The mode of consumption - whether individual or collective - was stressed as more important than Castells' general notion of 'collective consumption'. He argued that urban protests 'are more often defensive than aggressive' (Saunders, 1979:127) and they are thus not new challenges to the existing order.

The conclusion that Saunders came to on the 'new urban politics' of Castells was that it is both theoretically flawed and, on the basis of a comparison between Britain and France, largely inoperable in a British context.

The key theoretical chapter of <u>Urban Politics: a sociological</u>

approach concerns the question of the 'local state'. Saunders argued that increasingly directive and centralised state intervention has had important effects on the 'local state'. The latter has come to be less significant though it has retained 'a considerable degree of autonomy' from central government in respect of most of its functions. Saunders moves from a general discussion of theories of the state to a combination of (i) Offe's (1975) distinction between 'allocative' and 'productive' state activities and his analysis of technocratic crisis management strategies, and (ii) Winkler's (1976) corporatist thesis.

Saunders' general theory of the state is one that sees distinctive differences in functions at different spatial levels of state organisation, distinguishing between the 'corporate and non-corporate (or pluralist) sectors of state policy-making':

> The corporate sector would be concerned with production ... an increasingly independent and directive role ... The pluralist sector ... would be concerned mainly with consumption ... external to the various classes or political forces in civil society and thus open to influence by the most powerful of them (Saunders, 1979:178-179).

From that position, Saunders saw clear 'implications for the local state'. 'Production functions' (e.g. planning, road provision) require close relationships between local authorities and private business. 'Consumption functions' (e.g. education, social services, housing) operate in a more responsive millieu, so that 'we should expect the state responding to political pressures from the locality ...' (loc cit). In subsequent work Saunders has gone on to refine his position on the state further taking an explicit 'dual-state' stance (1981, Ch.8). Whilst arguing that the study of urban managers was of enduring importance, Saunders reviewed the question of local autonomy, concluding that the local state relies crucially on central state revenues from the private sector.

Dunleavy (1980) was critical of political scientists for focussing on 'local' policics or local government politics in lieu of 'urban' analysis and hence reflecting 'the ideological and institutional influence of government agencies, elected politicians and political parties' (Dunleavy, 1980:1). He stated his position firmly and clearly:

> I define urban politics as the study of decision processes involved in areas of collective consumption comprising primarily a series of public services, principally education, health care, environmental services, social welfare services, cultural facilities, highways, public transport, public housing, land use regulation and urban planning. This is essentially a content definition of urban politics (pp. 2-3, emphasis in the original).

He thus proposed a non-spatial definition of 'urban' firmly in the Castells' tradition scathingly dismissing other British work as pre-theoretical or non-theoretical. He rejected 'urban managerialism' as characterised by epistemological indeterminacy; moreover:

> The built-in focus on mediating institutions in urban managerialism ... generates a possibility of seriously misleading conclusions in which broader structural constraints and determinants of the local or regional context are lost to sight, and excessively individualistic and voluntaristic accounts of urban management are given on a one-off, non-cumulative basis (Dunleavy, 1980:42).

Dunleavy presented neo-marxist theories of state in a much more favourable manner, claiming that they no longer used a crude 'executive committee of the bourgeoisie' approach, rather they emphasised (i) the importance of state intervention in maintaining the long-run and general interests of capital, (ii) the existence of 'fractions' of capital or in some cases, 'sectors' of the economy with monopoly capital dominant and (iii) the role of the state in maintaining social cohesion and sustaining legitimacy.

Castells' (1977a, 1977b) arguments provided the basis for the development of Dunleavy's 'content definition of the urban field'. First, Dunleavy used the thesis that 'production activities are not fundamentally organised in terms or urban areas and cannot be used to explain the specificity of the urban' (p.45). Consumption, however, used to mean 'the extended (social) reproduction of labour power', he argued, gives the distinctive meaning to 'urban'. In making the distinction between individual consumption and collective consumption, too, he followed Castells. Individualised consumption was contrasted with collective consumption which equals those 'consumption processes whose organisation and management cannot be other than collective given the nature and size of the problem' (Castells, 1976, Dunleavy's emphasis). Such processes contrast with 'commodity forms of commercially organised consumption'. Dunleavy claimed that individualised forms of consumption are determined by class and income but that 'collective consumption processes can constitute an independent basis for the development of social cleavages'. Dunleavy thus defined urban politics 'in terms of the study of decision-making on collective consumption processes', a definition which, he argued, captures the most valuable work from spatial and institutional definitions (Dunleavy, 1980:45-50).

Dunleavy lists collective consumption processes - educational and health-care services, environmental facilities, highways, public transport, public housing, land use regulation and urban planning. The last two are 'not strictly concerned with consumption but with regulation or management of the urban system' (Dunleavy, 1980:53). They are included, however, as Castells has argued that they have major impacts on the development or collective consumption processes.

Having outlined an approach which derived almost entirely from Castells' work, Dunleavy drew back from full acceptance of the 'neo-marxist' theory on which Castells' view of urban processes was based, as he was 'highly sceptical of any attempts to deduce explanations of urban politics from the structural features of the capitalist mode of production' (Dunleavy, 1980:54). Following Lukes (1977) he maintained that structural explanations are neither causally adequate nor determinant accounts of social processes. He rejected the argument that each theoretical perspective _must_ constitute unique grounds for empirical assessment of its own propositions, and argued, in effect, for empirical analysis within a milieu of methodological pluralism.

For Saunders, then urban politics is about urban managers whereas for Dunleavy it is a study of political dimensions of policy areas. The positions that they adopt, however, and their assumptions do have much in common. Both are concerned with the nature of the state, and accept the separation of a largely autonomous 'political' sphere from an ill-defined sphere of 'economic' relations. They both equate, however modified, a concept of local state with the pre-given institutional parameters of British local government authorities. They both define 'urban' in terms of state practices, and they appeal to empirical validation of theoretical propositions.

These add up to _residual_ definitions of urban politics based fundamentally on particular aspects of British central-local government relations. Dunleavy's 'content field' relies on Castells' unproven assertions about _necessary_ state interventions and amounts to little more than a list of policy areas for focus. Why _these_ should be called 'urban' is beyond me. Dunleavy is much more convincing than Saunders, however, when he analyses extrnal influences on local practices, but entirely unconvincing when he claims a growing politicisation in terms of 'urban social movements'.

Given the very different inter-governmental structures and relations in Australia it is extraordinarily difficult to see how either Saunders' or Dunleavy's reformualtion of urban politics could be applied. The relationship between state governments and local government in Australia, to make the most obvious point, would require a complete reformulation of Dunleavy's 'content field' or Saunders' 'dual-state' hypothesis. Their work is thus effectively ethnocentric: theoretical generalisations rest crucially on the specifics of Britain in the 1970s. There is very little discussion in either work of the relationship between class and politics, and where it does occur, in Saunders' discussion of housing, it focusses on consumption categories rather than production relations. Nor is there much discussion of political parties and the class basis of formal politics.

Whilst both authors contribute valuable discussions of existing literatures their focus on Castells is at the cost of discussion of other developments in marxist theory which might inform an analysis of the state. Their theoretical contributions ultimately fail to resolve the specificity of the 'urban' dimension which they claim to operate within.

NON-CONTENTIOUS PERSPECTIVES ON LOCAL GOVERNMENT IN AUSTRALIA: POWER, WETTENHALL AND HALLIGAN

Local government systems in Australia is more of an encyclopedia than a mere book. As such I cannot attempt a comprehensive review - that task has to be left to others more familiar with local government in Australia. It is useful, however, to look at the issues covered in this book and in the light of these to reflect on the state of urban political analysis in Australia.

First we should note that the contributors were working to a tight schedule and to a brief which instructed them to describe the existing roles and functions of local government in Australia, but 'not in a contentious way' (p.x). The book comprises an editorial overview of local government in Australia and six papers on the local government systems of New South Wales, Victoria, Queensland, South Australia, Tasmania and Western Australia. The work was commissioned by the Advisory Council for Inter-governmental Relations (ACIR) and intended as a contribution to public policy formation. The contributors, then, include both academics and practitioners, with a strong practical orientation throughout. While the editorial overview raises many broader issues of theoretical importance and theother contributions also refer to wider issues, the book is not intended primarily as a theoretical contribution. It does, however, contain a mine of information on Australian local government and is essential reading for anyone interested in the subject.

The editorial overview is particularly useful. It demonstrates, first, historical variations in the development of local government between states, and suggests a broad dichotomy between 'constitutionalist' and 'state-interventionalist' local government systems. The former developed 'largely in response to local demand', (South Australia, Victoria) and the latter have been 'imposed from above' (New South Wales, Queensland). Subsequent chapters examine powers and responsibilities, types of local authorities, finance, inter-governmental relations and policy questions. The editors illustrate both general similarities and differences between and within the states by references to the companion papers on each state. In particular the authors of the introduction emphasise Bowman's insistence on variations in local government systems both within and especially between Australian states.

Fragmentation of responsibility and powers in Australian local government generally is a theme explored in chapter 3 'Local Authorities at Work'. Again, we should note metropolitan/non-metropolitan variations and variations between states, but most of all the lack of a unitary metropolitan (read 'urban'?) level of local government. With the one exception of Brisbane, local government in Australian cities is systematically and apparently irretrievably fragmented and disjointed. That is not, as the authors demonstrate, coincidental, rather it has been the outcome of specific relations articulating particular interests. Whilst some commentators bemoan the lack of unitary metropolitan authorities, Power, Wettenhall and Halligan demonstate convincingly why they have not emerged in Australia.

Local government finance is another important topic. Local government in Australia is not 'big business' as some commentators have described British local government. Although in absolute terms local government expenditure has increased, Chapter 4 notes that the financial role of local government in the economy 'has been small and gradually declining.' This contrasts, yet again, with the picture in the UK at least up to the mid 1970s, where local government expenditure was an increasing element of growing relative state expenditure.

Many other important questions are explored both in the introduction and the companion papers including innovation in local government, powers and responsibilities, and relations between levels of government. There is not, however, an extensive discussion of politics at the local government level, rather the concern is for 'inter-governmental relations' with some discussion of electoral systems and parties. Indeed, the focus of concern and emphasis throughout the book is on technical and administrative aspects of local government rather than on questions of party or class. In the paper by Bains and Mills on New South Wales, for example, there is no discussion of inner city party machine politics. The paper on Western Australia by Wood notes the increasing bureaucratisation of local government and suggests that the role of councillors has been correspondingly diminished - but does not consider the political implications of this. Tucker, writing on Queensland, has some discussion of the State government's opposition to Federal Labor encouragement of regionalism. The Queensland Premier announced in 1974 that the State Government would 'introduce legislation to prevent local authorities from dealing directly with the Commonwealth Government' (p. 546). However, discussion is couched in terms of the relationship between 'government' as if somehow governments had interests and roles separate from the political relationships contained within them.

Perhaps this perspective is inevitable given the nature of the brief to which the contributors were working, but the overall effect is to present a view of local government which is substantially apolitical, whatever theoretical perspective might be taken on defining politics. In the editors' introduction it is suggested that local government in Australia is largely not the domain of party politics, although we are told that this is not always the case:

> Non-partisanship appears to be a pervasive ethos in local government. It operates against candidates with

> affiliations or, where affiliations are used, favours local tickets rather than national political parties. The converse may be true in some metropolitan or large provincial municipalities, where tickets or parties dominate council elections and independents are generally unsuccessful (1981:33).

This suggestive comment, unfortunately, is not developed further so that readers interested in political aspects of local government in Australia may well come away rather frustrated. Indeed, as Parkin has recently argued (1979) it may be _precisely_ 'consensus' issues in Australian local politics that should be seen as problems _to be explained_. Nonetheless, there is an absolute mine of information in the book which will make it an essential reference book for all scholars of local government.

URBAN POLITICAL STUDIES IN AUSTRALIA

A review of three recent books hardly constitutes a discussion of 'the state of the art' in Australia - particularly when two of those books are British! Some critical problems have emerged, however, for attempts to develop urban politics as an arena of discourse in Australia. Dunleavy and Saunders both raised, but in my view failed to resolve the question of what a specifically 'urban' politics would constitute. Saunders' focus on 'urban managers' and his dual state thesis which specified particularly _local_ state activities rely heavily on the specifities of the British situation. Dunleavy emphasised the relationship between 'local' issues and practices and wider processes of state and market, but his reliance on Castells leaves him very exposed to the charge of functionalism.

The most crucial problems with attempts to apply either Saunders or Dunleavy to Australia then, are their ethnocentricity and lack of historical perspective. Their _theoretical_ propositions rely heavily on the organisation of central/non-central state relations in the UK and simply do not apply in the federal Australian context where, as Power, Wettenhall and Halligan demonstrate conclusively, local government is a very much _less_ important part of the state apparatus. The latter authors, however, do not attempt an anlysis of the _politics_ of local government in Australia and therby leave the question of urban political analysis unresolved.

Recent Australian work, however, has begun to focus on a range of issues and questions which, at first glance, might appear to fall appropriately under the rubric 'urban politics'. Parkin's work on city government, especially his recent book, focusses on the nature of Australian urban management emphasising the central role of the Australian States. Urban policy, too, attracted considerable attention during the 1970s and there is a growing literature on Federal initiatives towards Australian cities, for example Lloyd and Troy's (1982) study of the Department of Urban and Regional Development, _Innovation and Reaction_.

Different authors, however, use the term 'urban politics' to refer to diverse phenomena - Federal policies, local action groups, metropolitan planning, local government politics and so on. It is difficult to see a core or theme running through the diversity save that they refer to power, conflict and consensus over issues located in cities. That, however, is far too general a consideration to define a distinctive focus for research or teaching.

Few attempts have been made to incorporate Castells' work lock-stock-and-barrel into Australia. That is fortunate. As the debates and disagreements highlighted by a close reading of Saunders and Dunleavy show, Castells' contribution has been stimulating but it has also been extremely confusing! Having rejected the notion of an 'urban' sociology in The Urban Question he promptly invented a new 'urban sociology'. His manifold use of the term 'collective consumption' has generated yet more confusion - amongst his supporters as well as his critics. His conception of 'urban politics' and its significance is ultimately tautological and his assertion about 'urban social movements' and their significance are, simply, unsubstantiated.

From the implicit debate between Saunders and Dunleavy I can find no useful way to define 'urban politics' based on either the dual-state or collective consumption approach here in Australia.

The notion of a distinctively new and significant arena of struggle advocated by Castells wears very thin in Australian cities. At the same time, however, many of the issues discussed by Castells, Saunders and Dunleavy are important and worth studying. It may well be, in conclusion, that we can get on with studying them better if we abandon slippery concepts like 'urban politics' and focus instead on more conventional theoretical issues like class, power and the state.

REFERENCES

Carchedi, G. (1975) 'On the economic identification of the new middle class', Economy and Society, 4:1-86.

Castells, M. (1976) 'Theory and ideology in urban sociology', in Pickvance, C. ed., Urban Sociology: Critical Reading, London, Tavistock.

Castells, M. (1977a) The Urban Question, London, Edward Arnold.

Castells, M. (1977b) 'Towards a political urban sociology', in Harloe, M. ed., Captive Cities, London, Wiley.

Castells, M. (1978) City, Class and Power, London, Macmillan.

Dunleavy, P. (1980) Urban Political Analysis, London, Macmillan.

Lloyd, C. and Troy, P. (1981) Innovation and Reaction, Sydney, Allen and Unwin.

Lukes, S. (1977) Essays in Social Theory, London, Macmillan.

Offe, C. (1975) 'The Theory of the capitalist state and the problem of policy formation', in Lindberg, L. et. al. eds, Stress and Contradiction in Modern Capitalism, Lexington, Mass., Lexington Books.

Parkin, A. (1979) 'Cities without politics', Politics XIV, 2:291-

294.
Parkin, A. (1982) Governing the Cities: The Australian experience in perspective, Melbourne, Macmillan.
Power, J., Wettenhall, R. and Halligan, J. (1981) Local Government Systems in Australia, Canberra, AGPS.
Saunders, P. (1979) Urban Politics: A Sociological Interpretation Harmondsworth, Penguin edn.
Saunders, P. (1981) Social Theory and the Urban Question, London, Hutchinson.
Winkler, J. (1976) 'Corporatism', European Journal of Sociology, 17:100-136.

The view from the DURD bunker

Andrew Parkin

C.J. Lloyd and P.N. Troy, Innovation and Reaction : The life and death of the Federal Department of Urban and Regional Development, Allen and Unwin, Sydney, 1981, pp. 282, rrp $18.95 (case), $9.95 (paper).

Administrative history usually does not lend itself to exciting reading. The authors of this book on the genesis, life and demise of the Commonwealth Department of Urban and Regional Development (DURD) deserve credit, not least for sustaining a reasonably interesting narrative through 267 pages of often detailed description. Innovation and Reaction provides an insight into the problems engendered by the creation of a pretentious new Department, its developing relationships with other Departements and authorities, the administrative implications and complications of reformist public policy and the efficacy of more collegial organisational structures. As such, the book will feature conspicuously among texts on Australian public administration, though the onus is on the reader to plug this story into the general literature given the almost incomplete absence in the book of references to previous scholarship.

Innovation and Reaction, whether by intention or otherwise, seems less concerned with discussing urban policy per se as distinct from its bureaucratic implementation. There is a brief chapter outlining the Whitlam/Uren/Stretton/Butlin approach to urban policy which emerged in the 1960s and early 1970s (though the ANU Urban Research Unit's contribution is perhaps too modestly brushed over). There is also coverage of DURD's programmes and activities - land commissions, sewerage, area improvement, growth centres, urban rehabilitation, regionalism, and so on. However, the authors are generally content systematically to assert, rather than to demonstrate, the rationale for DURD's existence in the 1972-75 period. So we get bald, unsupported sentences such as "Analysis suggested that many urban problems could be solved if the process of development was approached in a different way" (157), judgements that particular DURD programmes were "successful" (for example, 180) without much discussion of what criteria for "success" might be, and a general sympathy for DURD's policy evangelism ("The task was to get the states and local governments to change their ways...", 158).

The book makes its contribution with the insights into the bureaucratic machinery. There are two aspects to this story - first, the internal arrangements within the Ministry of Urban and Regional Development and, second, the external bureaucratic relationships. The first aspect makes for the less interesting reading through detailed accounts of staff meetings, task forces, status levels, relations with brethren authorities (Cities Commission, National Capital Development Commission), Uren's role as Minister, and so on. Strangely, given that both authors were presumably non-civil service appointees to DURD, there is little discussion of the value of such "independent" advisers. It is also interesting that the innovatory staff pooling and task force arrangements attempted within DURD ran into difficulty with DURD's own staff as the career disadvantages of non-hierarchial arrangements became apparent.

External bureaucratic relations are given more lively treatment, possibly because the authors can be much surer about where their loyalties lie. Whereas DURD itself is portrayed as a complex aggregate of individual parts, multi-faceted and dynamic, the bureaucracy in general and other Departments in particular are portrayed as monolithic, unitary personalities capable of a consistency of motive and action. Clearly this account must be interpreted as simply the view from DURD's bunker. Lloyd and Troy would never tolerate for DURD itself the one-sentence characterisations which they apply to other Departments:

> [Other] departments didn't want to advise on how the government might best achieve its objectives, but were preoccupied with defending their territory - Housing and Construction was happy to support the proposition if it were given responsibility for constructing the offices.... Services and Property were distressed because of the implied criticism of its approach and because of a philosophical disagreement .... The [Public Service] Board was apprehensive. Treasury was ideologically opposed because it was evidence of yet another government intervention in the market place. Environment and Conservation was unhappy because environmental impact statements had not been conducted. Prime Minister and Cabinet was annoyed because it had not thought of the scheme and this was evidence that the <u>cordon sanitaire</u> it had erected around the Prime Minister could be penetrated (209).

Only occasionally is there (accidental?) revelation that DURD as viewed from outside might look remarkably like the outside world as viewed by DURD: "Treasurer Cairns ... commented to Uren that his officials had told him that the DURD representatives on the Officials Committee had supported all spending cuts - except those for DURD" (227). And consider the implications - Lloyd and Troy don't seem inclined to - of the following incident during the "constitutional crisis" of late 1975:

> Uren pointed to the hazards facing the Government and exhorted DURD officers to maintain the spirit which the Department had developed and to fight hard for urban and regional programmes which had been established, whatever happened to the Labor Government (236).

Does DURD differ from Treasury only in that such self-preserving sentiments are openly admitted? Plus ca change, plus c'est la même chose. In any event, the penultimate chapter rather sadly documents the transformation of DURD into the Department of Environment, Housing and Community Development and its subsequent gradual demise.

The authors' conclusions are modest but nevertheless somewhat optimistic: "reform can be carried out if a government is suitably tenacious and innorvatory. In particular DURD showed that whatever the weaknesses of the federal government system it can be made to work better by the patient development of cooperative policies" (:265). The DURD story as portrayed by Lloyd and Troy is a somewhat happier one for reformers than the usual accusations of trahison des clercs under Labor governments. Contrary to what (potential) reformist governments in Australia are probably now being advised, Lloyd and Troy suggest, though only implicitly, that a centralisation of administrative power is not necessarily desirable and that ministers and bureaucrats can work with a common purpose. Whether ministers can work cooperatively with each other is quite another matter. And I am not sure whether the narrative, as distinct from the conclusions, has really destroyed Martin Painter's alternative interpretation of DURD's ambitions and experiences (Australian Journal of Public Administration, December 1979) - that the whole thing was always impossible anyway.

Social mapping: a partial instrument

Nigel Thrift

Division of National Mapping Atlas of Population and Housing, 1976 Census. Volume 1: Perth, Canberra, 1979, 22 maps, rrp $12 (paper).
Division of National Mapping Atlas of Population and Housing, 1976 Census. Volume 2: Adelaide, Canberra, 1979, 27 maps, rrp $12 (paper).
Division of National Mapping Atlas of Population and Housing, 1976 Census. Volume 3: Brisbane and Gold Coast, Canberra, 1980, 27 maps, rrp $12 (paper).
G.T. McDonald and M.J. Guilfoyle Urban Social Atlas of Brisbane – 1976, Institute of Applied Social Research, Griffith University, Brisbane, 1981, pp. 207, rrp $10 (paper).
M. Poulsen and P. Spearritt Sydney. A Social and Political Atlas. Allen and Unwin, Sydney, 1981, pp. 164, rrp $49.95 (case), $29.95 (paper).

The 1976 Census has now given birth to a number of atlases. The volumes under review here form the majority of those already published (see also Houghton, 1979). Fortunately for their eyesight, most users of these atlases will probably only come into contact with but one or two of them and, certainly, they will not have to consider each and every map in detail. However the comparative opportunity such scrutiny affords does make it possible to assess the benefits and the limitations of each atlas, not just in terms of the variation in presentation and in the type of information conveyed, but also as a guide to the usefulness of so-called 'social mapping'.

Each of these atlases adopts a quite different solution to the visual presentation of census data on maps. The McDonald and Guilfoyle atlas of Brisbane is produced (and this is meant as no slur) at the lowest level of technical sophistication, in that SYMAP output is used which is then overlain with a base map and titles prepared by hand. The Poulsen and Spearritt atlas is produced via a CALCOMP pen plotter rather than the lineprinter and two transparent overlays are provided to help with the interpretation of each of the maps. Finally, the three Division of National Mapping atlases (another four are promised) are produced in colour, initially from computer output. Two fold-out base maps are appended at the end of each volume.

Particularly compact spatial distributions (e.g. Map 51, Hebrews, in the Sydney atlas) are always striking, so it is on the maps which have even more even distributions that each atlas's visual presentation must be judged. Given this criterion I would rank the Division of National Mapping atlases (with the advantage of colour) as the easist to comprehend visually, followed by the Poulsen and Spearritt atlas and the McDonald and Guilfoyle atlas. It is difficult to know how far the visual presentation of these atlases has been influenced by the wide and systematic map perception literature. My eyes tell me to suspect not much and this is a pity since this has been an area of considerable intellectual advance in the last few years.

Each of the atlases also uses different criteria for mapping the census data. Thus the McDonald and Guilfoyle atlas utilizes five class intervals based on percentages and presented at the suburbs level of spatial aggregation to produce maps of Brisbane. The Poulsen and Spearritt atlas of Sydney uses partially aggregated collection districts and five class intevals based on standard deviations. No mention is made of the modifiable areal unit problem, so one presumes this was taken into account. The Division of National Mapping atlases use collection district data and four equal class intervals based on persons per square kilometre. In terms of criteria used for mapping the Poulsen and Spearritt atlas is by far the most sophisticated. By using the standard deviation measure some of the more obvious statistical problems that the other atlases encounter are avoided; maps based on percentages or on equal classes can distort the spatial distribution of the populations under study in extreme cases, for instance.

Each atlas also varies in both the number and the subject of the maps. The McDonald and Guilfoyle atlas has 73 maps (72 if the base map is not included), the Poulsen and Spearritt atlas has 71 maps and the Division of National Mapping atlases have 20 maps, 25 maps and 25 maps again for the Perth, Adelaide and Brisbane and Gold Coast volumes respectively. These atlases vary considerably in terms of the variables used. The Division of National Mapping atlases use more general variables, the other atlases are more specific. In future it might well be worth considering the use of more ratio variables, mapped using a modified chi-squared statistic. This systematises the problem of selecting a limited number of variables for inclusion in an atlas and allows census information to be combined, often very effectively (see Census Research Unit, 1980). It is a more robust way of proceeding than, say, the factor analysis used is to produce four maps in the McDonald and Guilfoyle atlas of Brisbane.

In conclusion, I was left, after scanning these atlases, with one abiding impression. Leaving aside the Division of National Mapping atlases, which are pure atlases, I have to admit that I obtained far more from the text in both the McDonald and Guilfoyle and Poulsen and Spearritt atlases, than from the maps.

Indeed the title 'atlas' may be something of a misnomer in the case of these two volumes. The Poulsen and Spearritt volume, apart from the maps, also consist of a very enjoyable introduction - which is both a brief social and political history of Sydney and an enjoyable tilt at a subjects as diverse as high rents, the Australian Bureau of Statistics, sexism and political scientists - and a very detailed page of information (including tables listing the value of each variable by federal electorate, local government area and state electorate) accompanying each map. The McDonald and Guilfoyle atlas includes what are, in effect, short papers by different authors introducing each of the seven map sections of the Atlas. And each of these two volumes has an extensive bibliography.

Notwithstanding the obvious practical uses all of these atlases will have in attuning the delivery of services by local government and other bodies, in making it easier for advertising men to sell their wares and so on, this demonstrates once again to me that social mapping is, and will remain, only a very partial instrument for understanding spatial distributions.

REFERENCES

Census Research Unit (1980) People in Britain: A Census Atlas. HMSO, London.

Houghton, D.S. (1979) Perth at the 1976 Census: A Social Atlas. Department of Geography, University of Western Australia.

Books received

Bater, J.H. The Soviet City, Edward Arnold, London, 1980, pp. 196, rrp. $17.85 (paper).
Birrell, R. and Birrell, T. An Issue of People: Population and Australian Society, Longman Cheshire, Melbourne, 1981, pp. 277, rrp. $9.95 (paper).
Bowman, M. ed. Beyond the City: Case Studies in Community Structure and Development, Longman Cheshire, Melbourne, 1981, pp. 228, rrp. $9.95 (paper).
Brewer, F. On the Bread Line: Oral Records of Poverty, Brotherhood of St Lawrence and Hyland House, Melbourne, 1980, pp. 178, rrp. $9.95 (paper).
Burnley, I.H. The Australian Urban System, Longman Cheshire, Melbourne, 1980, pp. 339, rrp. $10.95 (paper).
Brown J. and Mullins, B. Town Life in Pioneer South Australia, Rigby, Adelaide, 1980, pp. 216, rrp. $14.95 (case).
Brunn, S.D. and Wheeler, J.O. eds., The American Metropolitan System: Present and Future, Edward Arnold, London, 1980, pp. 216, rrp. $37.50 (case).
Carter, H. The Study of Urban Geography, Edward Arnold, London, 3rd edition, 1981, pp. 434, rrp. $20.85 (paper).
Colwell, M. The Carroll Inheritance: A Novel about Colonial Adelaide, Rigby, Adelaide, 1980, pp. 174, rrp. $9.95 (case).
Daly, M.T. Sydney Boom Sydney Bust: The City and its Property Market 1850-1981, Allen and Unwin, Sydney, 1982, pp. 224, rrp. $9.95 (paper).
Evenson, N. Paris: A Century of Change 1878-1978, Yale University Press, New Haven, 1979, pp. 382, rrp. $29.95 (case).
Geismar L.L. and Geismar, S. Families in an Urban Mold: Policy Implications of an Australian-U.S. Comparison, Pergamon Press, New York, 1979, pp. 221, rrp. $36.00 (case).
Gilpin, A. The Australian Environment: 12 Controversial Issues, Sun Books, Melbourne, 1980, pp. 241, rrp. $8.95 (paper).
Hall, P. Great Planning Disasters, Penguin, Harmondsworth, 1981, pp. 308, rrp. $12.95 (paper).
Hanley, W. and Cooper, M. eds., Man and the Australian Environment, McGraw hill, Sydney, 1982, pp. 362, rrp. $18.25 (paper).
Herbert, D. The Geography of Urban Crime, Longman Cheshire, Melbourne, 1982, pp. 120, rrp. $13.90 (paper).
Johnson, R. The Green City, Macmillan, Melbourne, 1979, pp. 166, rrp. $12.95 (paper).
Kemeny, J. The Myth of Home Ownership, Routledge Kegan Paul, London, 1981, pp. 179, rrp. £6.95 (paper).
Knox, P. Urban Social Geography, Longman Cheshire, Melbourne, 1982, pp. 243, rrp. $22.30 (paper).
Logan, M.I., Whitelaw, J.S. and McKay, J. Urbanization: The Australian Experience, Shillington House, Melbourne, 1981, pp. 128, rrp. $9.50 (paper).
Logan, T. Urban and Regional Planning in Victoria, Shillington House, Melbourne, 1981, pp. 128, rrp. $9.50 (paper).

Mundey, J. Green Bans and Beyond, Angus and Robertson, Sydney, 1981, pp. 154, rrp. $8.95 (paper).

Nash, C.A. Economics of Public Transport, Longman Cheshire, Melbourne, 1982, pp. 194, rrp. $19.50 (paper).

National Library of Australia, Post-War Reconstruction in Australia: A Bibliography, the Library, Canberra, 1981, pp. 105, rrp. $6.50 (paper).

Neutze, M. Urban Development in Australia, Allen and Unwin, Sydney, 2nd edition, 1982, pp. 259, rrp. $13.95 (paper).

Parkin, A. and Patience A. eds., The Dunstan Decade: Social Democracy at the State Level, Longman Cheshire, Melbourne, 1981, pp. 310, rrp. $15.95 (paper).

Reinecke, I. Micro Invaders, Penguin, Harmondsworth, 1982, pp. 272, rrp. $6.95 (paper).

Roberts, K. Leisure, Longman Cheshire, Melbourne, 1982, pp. 140, rrp. $11.00 (paper).

Roe, J. ed. Twentieth Century Sydney: Essays in Urban History, Hale and Iremonger, Sydney, 1980, pp. 272, rrp. $14.95 (paper), $24.95 (case).

Scott, R. ed. Interest Groups and Public Policy: Case Studies from the Australian States, Macmillan, Melbourne, 1980, pp. 248, rrp. $10.95 (paper).

Solomon, A.P. ed. The Prospective City, MIT Press, Cambridge, 1980, pp. 491, rrp. $13.95 (paper).

Spearritt, P. The Sydney Harbour Bridge: A Life, Allen and Unwin, Sydney, 1982, pp. 120, rrp. $9.95 (paper), $15.95 (case).

Troy, P.N. ed. Equity in the City, Allen and Unwin, Sydney, 1981, pp. 229, rrp. $8.95 (paper), $19.95 (case).

Troy, P.N. ed. A Just Society? Essays on Equity in Australia, Allen and Unwin, Sydney, 1981, pp. 247, rrp. $8.95 (paper), $19.95 (case).

Weinstein, M.S. Health in the City, Pergamon Press, New York, 1980, pp. 92, rrp. $15.00 (case).

Yeates, M. North American Urban Patterns, Edward Arnold, London, 1980, pp. 168, rrp. $32.85 (case).

NOTE

Listing here does not preclude review in a subsequent issue of the Yearbook.

For Product Safety Concerns and Information please contact our EU representative GPSR@taylorandfrancis.com
Taylor & Francis Verlag GmbH, Kaufingerstraße 24, 80331 München, Germany

www.ingramcontent.com/pod-product-compliance
Lightning Source LLC
LaVergne TN
LVHW011126300926
849160LV00008B/533

* 9 7 8 0 4 1 5 8 6 0 3 7 6 *